Technology for Children
Developing your own approach

Marilyn Fleer & Beverley Jane

PRENTICE HALL

Sydney • New York • Toronto • New Delhi
London • Tokyo • Singapore • Rio de Janeiro

© 1999 by Prentice Hall Australia

All rights reserved. No part of this publication may be reproduced, stored in a retrieval system, or transmitted in any form or by any means, electronic, mechanical, photocopying, recording or otherwise, without the written permission of the publisher.

Acquisitions Editor: Lachlan McMahon
Production Editor: Elizabeth Thomas
Text design: Jack Jagtenberg
Typeset by Southern Star Design, NSW
Printed in Australia by Star Printery, Erskineville, NSW

1 2 3 4 5 03 02 01 00 99

ISBN 0 7248 1210 5

National Library of Australia
Cataloguing-in-Publication Data

Fleer, Marilyn.
 Technology for children: developing your own approach.

 Bibliography.
 Includes index.
 ISBN 0 7248 1210 5.

 1. Technology - Study and teaching (Primary). I. Jane, Beverley. II. Title.

372.358044

Prentice Hall of Australia Pty Ltd, *Sydney*
Prentice Hall, Inc., *Upper Saddle River, New Jersey*
Prentice Hall Canada, Inc., *Toronto*
Prentice Hall Hispanoamericana, *SA, Mexico*
Prentice Hall of India Private Ltd, *New Delhi*
Prentice Hall International, Inc., *London*
Prentice Hall of Japan, Inc., *Tokyo*
Simon & Schuster (Asia) Pte Ltd
Editora Prentice Hall do Brasil Ltda, *Rio de Janeiro*

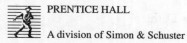

PRENTICE HALL
A division of Simon & Schuster

Contents

Preface v

PART 1 **THINKING ABOUT TECHNOLOGY** 1

CHAPTER 1 *Our experiences and understandings of technology and technology teaching* 3

CHAPTER 2 *Technology in the home: children's experiences of technology* 23

CHAPTER 3 *Multiple world views in curriculum design and implementation: cultural construction of technology* 37

PART 2 **APPROACHES TO TEACHING TECHNOLOGY EDUCATION** 55

CHAPTER 4 *Science—technology relationship* 57

CHAPTER 5 *A symbiotic approach* 81

CHAPTER 6 *The process approach to teaching technology* 95

CHAPTER 7 *An ecological approach: using technology appropriately* 111

CHAPTER 8 *Developing your own approach to technology education* 123

PART 3 **PLANNING FOR TECHNOLOGICAL LEARNING** 137

CHAPTER 9 *Finding out children's technological capabilities* 139

CHAPTER 10 *A question of design* 153

CHAPTER 11 *Discourses in technology education* 161

CHAPTER 12 *Cooperative technological learning* 177

CHAPTER 13 *Conclusion: what is the magic ingredient?* 191

Index 202

PREFACE

A book of research

Technology education is a new area of study. Educationalists are only now beginning to understand how best to develop programs in this area, and how to effectively work with children to develop their technological capabilities.

The content of this book is based upon research and practice in technology education with Australian children and pre-service teachers. Much of this research has been supported by grants from the University of Canberra, Australian Research Council Small Grant Scheme and the Curriculum Corporation of Australia. Data from both authors' PhD theses have also been used to support discussions contained within this book. As a result, this book represents a comprehensive account of research into technology education within Australia. The book seeks to make a scholarly contribution to knowledge construction in technology education within Australia.

Design features of the book

The main design feature of this book is interactivity. Readers are encouraged to reflect upon the research reported in each of the chapters and to consider what their position may be in relation to the vignettes detailed. The first part of the book encourages readers to think broadly about their definition of technology and technology education. It is argued that the more broadly technology is conceived, the more likely teaching programs will be diverse and socially relevant for children. Through this re-interpretation of the literature, critical insights into technology education are possible.

The second part of this book presents a range of approaches to the teaching of technology education. The theories of teaching and learning in technology education are systematically critiqued. Examples of design briefs, which draw upon a particular approach, are included in each chapter in the second part of the book. Readers are invited to reflect upon each approach and make decisions about the possible learning outcomes for children. In Chapter 8, readers are asked to decide upon their own approach to teaching technology education.

The third part of the book provides a contemporary analysis of planning for technology teaching. It is argued that learning can be maximised for children when teachers provide an environment in which:

- children have a sense of purpose for working;
- teachers organise opportunities for observing children's technological capabilities;
- teachers utilise observations as the basis for planning;
- children use their technological constructions/processes (rather than simply making and then taking them home);

- cooperative learning is not only fostered but taught as an important skill in working technologically;
- teachers work towards helping children to ask questions of their designed environment; and
- teachers critique the range of discourses that emerge and ensure that all children have agency.

Acknowledgments

Developments in scholarly knowledge of technology education are only possible when researchers have the space in their lives to analyse, reflect and apply intellectual energy to research. We wish to acknowledge the support of our families and the universities (particularly Deakin University) within which we work. Freya Fleer-Stout and Rowan Fleer-Stout have provided many opportunities for one of the authors to observe and reflect upon technological activity they engaged in, noting many contradictions with the literature. Similarly, research funds, acknowledged earlier, greatly assisted us with the task of developing supporting material for the arguments presented in this book.

Many teachers and their children participated in the research we undertook in writing this book. Individuals have been acknowledged in the relevant chapters. Special acknowledgment is made of Anna Howell and Alan Nicol from the University of Canberra, who videotaped many hours of teaching and edited large quantities of data, which were used to support several chapters in this book.

Finally, special acknowledgment is made of all the pre-service teachers who participated in the research, which formed the basis of a number of chapters within this book.

MARILYN FLEER AND BEVERLEY JANE

Part 1

Thinking about technology

Chapter One

Our experiences and understandings of technology and technology teaching

INTRODUCTION

The term 'technology' is heard in many different forums today. There are many programs devoted to and reporting on the latest technological inventions. Newspapers often feature articles associated with technology. Many politicians associate economic success with technological products and capabilities. Yet what do we really mean when we talk about technology? This chapter is the beginning of a journey into exploring your own thinking, learning and teaching of technology. In developing an understanding of what technology is and what technology education may look like, we suggest you keep a journal.

> JOURNAL ENTRY 1.1   *Recording your journey*
> Before you proceed, obtain a book with bound pages so that you can keep a cumulative record of your thoughts as you progress through this book. It is our belief that your ideas will develop as you read and reflect on the contents of this book.
>
>
>
> Figure 1.1   *Writing our ideas in our journal*

Your responses to the activities listed throughout the book should also be recorded in your journal as they will influence your thinking. It is important to complete the activities listed as you encounter them. Do not read on unless you have at least thought about your views on the questions raised. The text that follows will be more meaningful and provide more challenge for you if you work in this way.

> JOURNAL ENTRY 1.2 *What is technology?*
>
> In your journal record your views on 'What is technology?'.
>   What are your images, concepts and feelings about technology in general and how does technology fit in with your life and society?

It is likely that the views you have recorded will look similar to those shown below in Figure 1.2.

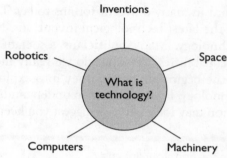

Figure 1.2   Views of technology

Figure 1.2 shows the group views from a class of undergraduate students undertaking a science and technology education unit. Their views indicate many of the 'high' technologies that currently exist within our society. What other things could have been included? Ask your friends what comes to mind when they think of technology. Note whether or not their responses are similar or different. Further examples of people's views on technology are shown in Figures 1.3 and 1.4.

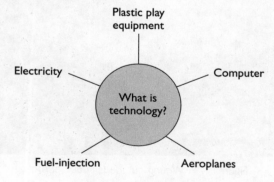

Figure 1.3   Views of technology

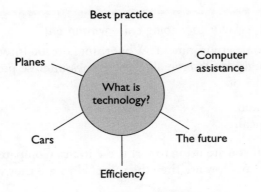

Figure 1.4 Views of technology

In the book *Science for Children* an activity termed 'Which is technology?' is listed (Fleer & Hardy 1996, pp. 168–69). This activity modified from Symington (1987) is useful for clarifying your thinking further.

JOURNAL ENTRY 1.3 *Which is technology?*

Do you associate the following in your mind with technology? Put a tick (√) against the item if you **do**, and a cross (X) if you **do not**. (25 items)

Ballpoint pen	Newspaper	Brick
Packet of coffee	Bulldog	Pair of spectacles
Clothes peg	Racehorse	Computer
Radio	Robot	Seedless tomato
Soap	Hammer	Hairdryer
Corkscrew	Iceblock stick	Kidney transplant
Microwave oven	Laser	New variety of wheat
TV	Test-tube baby	Toaster
Woollen dress		

Those people who have done this activity have tended to associate technology with an object that has been introduced into our community recently, such as a computer or laser. However, other forms of technology such as a microwave oven, gas cook top or fire were new technologies at some point in history! An iceblock stick is not often thought of as technology, nor a paper clip or digging stick. These are simple technologies which are equally as important as high technologies in society. Cultivating a new variety of wheat or a seedless tomato is a technological process which also comes under the banner of technology. There are many processes and products which can be considered as technology.

JOURNAL ENTRY 1.4 *Our technological environment*

Look around your home environment. What things do you interact with daily that you believe are technological? Observe all those things in close proximity to you and list them in the categories of:

- technological; and
- non-technological.

Identify the criterion you are using to sort these items. Compare your views with those presented in Figure 1.6. How are they the same? How are they different? What do you notice?

Figure 1.5 Caring life support technologies are important for the well being of humans

A large portion of our environment is constructed. How does our environment influence how we interact? What parts of our environment could be considered as technological? Do children have the same view of technology as many adults?

Young children have also been asked about their views on technology. Jarvis and Rennie (1994) used a picture quiz to identify Australian and English children's perceptions of the term 'technology'. They asked children (aged seven and eleven) to identify from a list of 28 items (shown below) which items had something to do with technology.

Computer	Advertisement	Microwave oven
Statue	Telephone	Sheet of music
Aeroplane	Clock	Stone Age axe
Factory	Windmill	Gum tree/oak tree
Volcano	Cup	Jeans
Rose	Platypus/fox	Cough medicine
Poodle	Cheese	Fish and chip shop
Gun	Plan of house	Mine
Bridge	Bedroom	
Book	Playground	

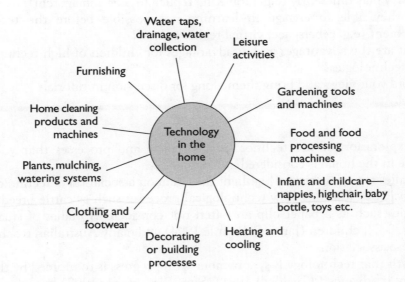

Figure 1.6 *Technology in the home*

These children responded in a similar way to those in Symington's (1987) study. The most frequently listed item was the computer (high technology) and the least frequent was the poodle (technological process).

How will a child's environment be different to your own? Imagine you are in a young child's house. (This may not be too difficult for those of you who have children yourselves.) Young children have access to many 'high' technological toys, games, experiences and equipment. The range is quite staggering:

- dolls with a voice chip (for talking)
- remote-controlled vehicles
- books with music or sound chips
- CD Roms designed for children
- Computerised games
- Videos

- Cassettes
- Soft toys with computer-controlled movements.

The range of technological artefacts available to children has escalated over the last ten years. Technological products have improved and become cheaper, thus making them more readily accessible to most families.

> JOURNAL ENTRY 1.5 *Children's technological worlds*
>
> In what way do the recently introduced technological items influence children?
> Do they play differently (e.g. Lion King replica toys or Tamagotchi)?
> Are they able to engage in learning not possible before the technological development (e.g. cyberspace, virtual reality games)?
> What are the advantages and disadvantages for children of high technologies and simple technologies?
> Record your views and bring them along for discussion in tutorials.

Further exploration of the technological artefacts and processes that young children experience in the home is considered in Chapter 2.

The studies reported thus far highlight the common association of technology with high technology. They illustrate how technological processes such as cattle breeding or simple technologies such as a paper clip are often not considered. Studies of student teachers (Hardy 1990–7), children (Jarvis & Rennie 1994) and many Australian teachers have also shown these associations.

The myth that technology is synonymous with progress is reinforced by the belief that technology is advanced (Turnbull 1991). Siraj-Blatchford (1997) has argued that much of the popular literature presents a 'linear continuum between high and low technology, inferior and superior cultures, primitive and advanced civilisations' (p. 38). He also states:

> *Technology for most people means 'high' technology, 'labour saving' technology, 'high consumption' technology, and they also see the use of such technologies as a measure of social and cultural development* (p. 37).

There has been a strong association between technology and advancement in Western cultures. Siraj-Blatchford has argued that:

> *Historically the use of slave labour and women's oppression have undoubtably been the most significant factor in explaining the development of Western science and technology* (p. 74).

Siraj-Blatchford challenges readers to carefully think through their assumptions, values and perspectives on technology and technology education.

In an important study investigating adult perceptions of technology Hardy (1997) found that there are multiple representations of what is considered as technology. In the analysis of the discourse of the four women he interviewed he says:

- the women find it meaningful to discuss their general conceptions of science and of technology, which in most instances are stable and applied consistently to specific contexts in their lives;
- among the four women there are marked differences in these conceptions;
- technology is typically spoken of as something immediate, is often seen as embodied in machinery and appliances, is related to some notion of advance, but in specific areas was critiqued, and induces relatively strong feelings;
- the women readily nominate technologies as segregated by gender (with most domestic technologies generally the province of women);
- the women differ on the extent to which they accept the gender divisions in the use of technology, and there is evidence that for some there are elements of struggle to gain more scientific and technological knowledge and control in specific areas of importance to their everyday lives (A section taken from Hardy 1997, pp. 24–5).

Hardy also found that two of the women interviewed had beliefs in relation to ageing and Christianity which influenced their thinking about technology. He suggested that it is important in understanding people's experiences of science and technology to consider:

- the way that discourses on ageing can create future or past orientations to development and the value of science and technology; and
- the way that discourses on religious beliefs can create supportive or critical orientations to developments and the value of science and technology (p. 25).

Hardy believes that 'it is critical to understand more adequately the complexity and diversity of their gendered and scientifically and technologically textured life experiences. Educators and communicators might then engage more actively with the ideas of women building on their understanding of, and wisdom about, the functioning of science and technology' (p. 26).

It is interesting to note that of the technological artefacts or processes detailed in Figures 1.2 to 1.4, technologies often used by women were not included (e.g life-support technologies such as a cook top, fridge or vacuum cleaner; caring technologies, such as a baby bottle, nappy or dummy). Analyse the responses in journal entry 1.2 in relation to gender. Further thought is given to gender and technology in Chapter 11.

TECHNOLOGICAL HISTORY

In Figure 1.7 a hot-air engine for opening temple doors is shown. According to Siraj-Blatchford (1997) many modern day inventions have been attributed to Western cultures. Siraj-Blatchford has argued that cultural chauvinism runs deep and Eurocentrism has a very long history (p. 26). For example:

- While many contemporary history books cite Archimedes (287–212 BC) as the inventor of the lever, a wide range of simple machines must have been employed by Imhotep (2650 BC) to build the first of the Egyptian pyramids (p. 27);

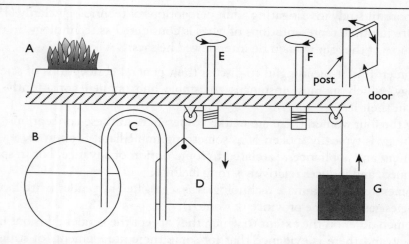

When the fire on the hollow alter A was lit, the air within it expanded and pushed the water out of the hidden tank B. The water passed through the pipe C and into the hanging bucket D which then dropped causing the vertical door posts E and F to rotate. When the fire was extinguished the air in the altar contracted and atmospheric pressure caused the water to pass back from the bucket to the tank (D to B). At this point the counter weight G dropped to cause the door posts to turn in the opposite direction and close the doors.

Source: Siraj-Blatchford, 1997. Reproduced with permission.

(Heron of Alexandria—1st Century AD)

Figure 1.7 A hot-air engine for opening temple doors

- The invention of the electric lamp which is attributed to American Thomas Edison (1879) was patented by Lewis Latimer who developed the practical lamp, including a filament that made electric lighting a real alternative to gas. Latimer was an Afro-American.
- A 7th century Islamic vertical axis windmill and pump are described in Ibn al-Razzaz al-Jazari's *Book of Knowledge of Ingenious Mechanical Devices*. The pump converts rotary to reciprocating motion (cited in Siraj-Blatchford 1997).

Siraj-Blatchford has suggested that a Eurocentric approach to the labelling and hence recording of invention of technological artefacts and processes reflects a culturally chauvinistic attitude. [Hence racism can be reflected in aspects of design and technology.] Although an anti-racist approach should be used in curriculum content selection, encouraging children to adopt this perspective when examining historical accounts is more difficult to engineer. For example, Siraj-Blatchford has suggested that curriculum material published through the education authorities in England encourages children to evaluate technologies from other cultures. A letter from a Ghanian village is presented in one of the suggested units of work. This letter requests charity to help establish a water supply system. The unit is organised to encourage British children to decide upon how the villagers are to be assisted. Siraj-Blatchford argues that:

> *This is an exercise directing European children to evaluate the technological choices for rural Africans and while a great deal of information is provided about the science and*

technology, [no information is provided about the culture]. This is the worst possible form of 'intermediate technology' education, where the charitable minority world scientists and technologists provide inappropriate technological solutions to the dependent and (at least implicitly) grateful majority world communities (p. 38).

[As educators we need to foster respect for the multiple ways people have developed solutions to problems.] The focus must be on cultural similarities and shared needs. As educators we must move away from using examples from the poor majority (often referred to as developing countries or third world countries) to the over-consumption of the rich minority. More thought is given to this in Chapter 3.

THE SOCIAL SHAPING OF TECHNOLOGY

> ... *it is no good studying technology separately from the context in which it is used and abused* (Siraj-Blatchford 1997, p. 33).

It has been argued elsewhere (Fleer & Hardy 1996) that many suppose that technological artefacts create social change. (For a detailed discussion of technological determinism see MacKenzie & Wajcman 1985.) For example, many people thought that the introduction of computers would reduce the number of jobs available. [The attention was directed towards how computers would create change, rather than examining the social context which brought about the introduction of computers.] Similarly, workers voiced concern during the industrial revolution regarding the introduction of the power loom, arguing against the use of machinery, since jobs would be lost. They did not question the social change that was occurring through bringing together workers under one roof, whereby more control could be exercised over the workers (how they worked, when they worked, what they could produce—the loss of independence and change in social structure as people moved from their home to a work site).

Technology emerges from within a social context. The following example illustrates how the technology of obstetrics dominated by males emerged within a social context in which women were the healers and midwives—they predominantly held this scientific and technological knowledge. As you read note how the social context was the catalyst for the technological change. [An elaborate example is necessary for the effective demonstration of how the social factors operating at the time brought about the development of a technological artefact in midwifery.] Please note that the early writings in Revisionist Feminist Projects (e.g. Oakley 1976) have established many understandings that still continue to be used today. Some earlier important original works have been drawn upon and cited throughout the following discussion (e.g. Ehrenreich & English 1973; Haire 1973; Oakley 1976; Kitzinger & Davis 1978).

Childbirth and technology

Until the 17th century support for childbirth had been the preserve of women (Wajcman 1991). It was midwives who attended and managed labour in the home (Oakley 1976). Their knowledge and skill about birth and birthing were passed on from one generation to the next (Wajcman 1991), principally through an apprenticeship (Oakley 1976). Women were seen as the healers, the unlicensed doctors and anatomists of Western history (Ehrenreich & English 1973). Yet today pregnancy and childbirth for most women is enshrined in a carefully structured, sequenced and managed medical program, predominantly in institutions under the direction of medical practitioners (most of whom are men). Over the course of three centuries childbirth has become medicalised (Ehrenreich & English 1973) with devastating effects upon women and their babies, socially (Wajcman 1991), emotionally (Ehrenreich & English 1973; Kitzinger & Davis, 1978), politically and physically (Haire 1973).

> JOURNAL ENTRY 1.6 *Is there a link between technology and childbirth?*
> Record your views.

Historically there are two periods in which the control of childbirth moved from women to men (Oakley 1976). First from the 14th to the 17th centuries in Europe where midwives and female lay healers were suppressed, and again in the 19th and 20th centuries where obstetrics was included in the curricula of professional medical training.

During the period between the 14th and the 17th centuries calculated and explicit suppression of women healers (including midwives) took place. This important scientific and technological knowledge held by women was used extensively, particularly with the poor. The violent exclusion of women as independent healers was directed towards the working class. Ehrenreich and English (1973) have outlined this persecution through the documentation of the 'witch hunt craze':

> *The witch-craze took different forms at different times and places, but never lost its essential character: that of a ruling class campaign of terror directed against the female peasant population. Witches represented a political, religious and sexual threat to the Protestant and Catholic churches alike, as well as to the state* (p. 7).

The empirical approach by the female lay healers contrasts also with the anti-empirical and Church doctrine focused approach taken by universities, where medical training was just beginning (14th to 17th centuries). The distinction in practice can be shown through the writings of Ehrenreich and English (1973):

> *Confronted with a sick person, the university-trained physician had little to go on but superstition. Bleeding was a common practice, especially in the case of wounds. Leeches were applied according to the time, the hour, the air, and other similar considerations. Medical theories were often grounded more in 'logic' than in observation: Some foods brought on good humours, and others, evil humours ...*
>
> *Such was the state of medical 'science' at the time when witch-healers were persecuted for being practitioners of 'magic'. It was witches who developed an extensive understanding of bones and muscles, herbs and drugs, while physicians were still deriving their prognoses from astrology and alchemists were trying to turn lead into gold ...* (Ehrenreich & English 1973, pp. 16–17).

> **JOURNAL ENTRY 1.7** *Analysing the social context*
> List the range of technological knowledge held by female midwives and healers.
>   How was this knowledge gained?
>   Was this knowledge valued by society?
>   Consider the changing social context surrounding midwifery and healing that is discussed below. As you read list the social forces operating.

problematic & excluded from uni.

* [It was through the institutionalisation of science, specifically medicine that the way was paved for excluding women from practising healing and midwifery.] Women were excluded from universities. They were barred from attendance and hence could never legally practise medicine (Ehrenreich & English 1973). Competition between university-trained doctors and lay healers, and later midwives, saw pressure being placed upon governments to legislate against the practising of medicine without qualifications. Unfortunately, few trained physicians were available to treat the masses, particularly the poor who could least afford expensive doctors' fees (Ehrenreich & English 1973; Kitzinger & Davis 1978; Wajcman, 1991).

[As the public image of medicine as a profession grew, campaigns were run to label midwives as 'dirty', 'incompetent' and 'uneducated'. Ehrenreich and English (1978) cite how medical journals encouraged support for this campaign:]

> ... surely we have enough influence and friends to procure the needed legislation. Make yourselves heard in the land; and the ignorant meddlesome midwife will soon be a thing of the past (Ehrenreich & English 1978, p. 97).

* Similarly, in Australia women were made to feel inferior if they did not elect to use the services of a doctor for childbirth:

> Medical care was regarded [by women in rural communities] often as a non-essential luxury, although occasionally an expectant mother took lodgings in a town where it was available. However, 'the majority of bush women preferred to stay at home and make shift with the peripatetic Gamp, old and unscientific as she was', or with just a woman neighbour to help her (Forster 1967, p. 14).

The emerging specialisation within medicine of gynaecology and obstetrics had the promise of great financial gains. Consequently, midwifery provided direct competition, first with the upper class and later within the lower classes. Direct intervention was becoming necessary. Even within Australia such discussions were documented:

> Later, doctor pre-eminence in obstetrics was continued by the medical profession's failure to provide any systematic training of midwives. There were heated arguments over what form training, if instituted, should assume, because many doctors feared that the fully qualified midwife would not only take over obstetrical practice, but also invite the lucrative field of diseases of women (Forster 1967, p. 15).

The continual discrediting of midwifery practice by the emerging medical profession had variable effect. In America midwifery was eventually eliminated (Ehrenreich & English 1978) and in England expensive training and registration became necessary (Oakley 1976). Thus institutionalising the control of midwifery from women to the male-dominated medical profession.

Another important factor in the social control of childbirth has been the move towards hospital childbirth. This change in practice was initiated by the medical profession. Oakley (1976) writes:

> ... the hospitalisation policy evolved over the period from 1910 to about 1940 in Britain. In the early 1900s it was impossible to persuade a 'respectable' woman to have her baby in hospital. After 1944, the official policy in Britain was a 70% hospital confinement rate (p. 48).

The practice of hospital childbirth did two things to reinforce the medicalisation of childbirth. First it placed the women in an environment which ensured that the management of her childbirth was under the direct control of the male doctor, and second it sealed the definition of childbirth as an 'illness' to be treated by the medical profession. The resultant anxiety in the expectant mother can be summed up by Davidson and Rakusen (1982):

> It is a clear example of how the medical profession can fail to see women as intelligent individuals to be treated with respect and dignity and to be given information as a right. A clue to the reason for this gap between what women actually want and what the medical worker allows them to have is suggested by midwife Jean Walker: 'the process of turning child-bearing women into patients for treatment in hospitals is unfortunate and may sometimes produce undesirable side effects. Indeed, ante-natal education is often geared as much to teaching people to be "good" patients as it is to helping them to be "good" parents.' (p. 33).

In this instance a 'good' patient is one who doesn't waste a doctor's time by asking questions, but rather one who is happily subjecting herself to an examination and to answering questions. Davidson and Rakusen (1982) cite the British Medical Association's family doctor booklet *You and Your Baby*:

> ... you are going to have to answer a lot of questions and be the subject of a lot of examinations. Never worry your head about any of these. They are necessary, they are in the interests of your baby and yourself and none of these will ever hurt you (p. 33).

The mind-body separation (which arose from the Cartesian dualism) provided a basis for constructing the body as a machine within medical ideology. Birthing women were not only seen as passive patients but as 'manipulable reproductive machines' (Oakley 1976, p. 65). This split was part of the 'natural' progression in thinking which linked female and male with irrational and rational, and which allowed the dualism between midwifery and medicine to strengthen.

Over time the gradual positioning of midwifery as inferior to university-trained physicians in childbirth occurred. Yet most of the claims made by the male medical profession to discredit midwifery practice were unfounded. Llewellyn-Jones (1979) has documented significant evidence to support the notion that university-trained doctors actually increased maternal morbidity. Similarly, Oakley (1976) cites an example of antisepsis in 1847:

> *Semmelweiss in Vienna observed that the death rate of mothers in hospital wards attended by medical students was three times that in wards attended by midwives. His correct explanation of this was that the medical students brought contagion with them from the post-mortem room (some of the higher incidence was probably also due to lack of hygiene in the medical students' use of forceps)* (p. 35).

It was not just the incompetence or unhygienic practices of the medical profession that were of concern, but the de-humanisation of the birthing context. Oakley (1976) writes:

> *... obstetricians introduced new dangers into the process of childbirth. Unlike a midwife, a doctor was not about to sit around for hours, as one doctor put it, 'watching a hole'; if the labour was going too slow for his schedule he intervened with knife or forceps, often to the detriment of the mother or child ... the physicians were usually less experienced than midwives, less observant, and less likely to even be present at a critical moment* (p. 98).

> JOURNAL ENTRY 1.8 *The invention of forceps*
>
> In your view what were the most pertinent social conditions operating at the time, which facilitated the invention and widespread use of forceps?
> Record your views.

The development of technological artefacts played a significant role in the emerging male-dominated medical profession and their take-over of childbirth. One example of a technology which legitimated male involvement in the birthing process was forceps. Obstetric forceps introduced in the 1700s by the Chamberlen family, according to Oakley (1976) or through Scottish apothecary William Smellie as documented by Wajcman (1991), allowed a division between male and female midwifery practice to occur. Since custom precluded midwives from using instruments as an accepted part of their practice (Wajcman 1991), male midwives seized this opportunity to provide a form of specialist assistance during childbirth.

The male midwives aligned themselves with the more prestigious specialisation within medicine of surgery (originally as members of the barber-surgeon guilds) (Oakley 1976). The use of instruments had already become the reserved right for their exclusive (male) membership, hence the alliance through their use of forceps seemed natural. Initially surgeons had been called upon to assist with difficult deliveries, later this became the preserve of the male midwives.

At this time fashion also began dictating to the upper class that it was more desirable to have a male midwife in attendance at childbirth (Oakley 1976). Consequently, the outcome was for an increase in the use of technology and for the rate of intervention during childbirth to escalate. Male midwifery could now be clearly distinguished by its use of technological artefacts (Wajcman 1991).

The use of technology not only had devastating effects upon women (due to internal damage) and their babies (frequently death) due to premature intervention, but it changed the relationship between the woman and the physician. The Cartesian model of the body as a machine and the physician as a technician or mechanic was reinforced through the use of technological artefacts (Wajcman 1991, p. 67). Here the beginnings of the oppression of women through the loss of control of their bodies could be explicitly seen. The physician with obstetrical equipment and knowledge knew best (Ehrenreich & English 1978). Oakley's (1976) account of the Chamberlen family blindfolding women during childbirth when forceps were being used to safeguard the secrecy of the technical equipment is a case in point. Modern obstetricians have continued this mystique through the use of the stethoscope and foetal monitoring machines. In this way they have positioned themselves to give the appearance of knowing more about women's bodies than the women themselves (Wajcman 1991). Interestingly though this technology (stethoscope) was not developed because of any technical deficiency on the part of the ear, but because of prevailing social mores (Wajcman 1991, p. 70). Wajcman (1991) explains:

> *The ubiquitous stethoscope has its origins in the doctor's wish to keep the patient at a distance, overlaid with the requirements of modesty as between men and women* (p. 70).

JOURNAL ENTRY 1.9 *Artefacts facilitating childbirth*
What other artefacts have emerged to facilitate childbirth? Record your views.

Ultrasound imaging in pregnancy can be viewed as the obstetrician taking control of pregnancy and knowing more about the foetus' development than the expectant mother herself.

> *The procedure serves to discredit and then displace women's own experience of the progress of the foetus in favour of scientific data on the monitor* (Wajcman 1991, p. 71).

In Western medicine the use of high technology and technological activity are accorded not only great status but huge sums of money. The further specialisation within reproductive science, notably IVF, has provided yet another lucrative industry with high status at the expense of women (Wajcman 1991). However, currently the financial rewards in obstetrics are high and as a consequence great resistance to changing current practices and beliefs is likely. The social shaping of technology will continue whilst the rewards are high and there is status associated with its use (MacKenzie & Wajcman 1985). Midwifery continues to hold an inferior position since the knowledge and skill held are not overtly linked with a technological artefact.

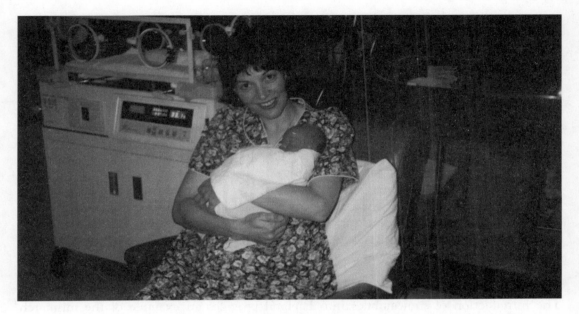

Figure 1.8 The arrival of a new baby: midwifery is not often considered a technology because it is a system of knowing and doing rather than a tangible artefact

Social scientists have made much of the social effects of technology and its impact on society (technological determinism: MacKenzie & Wajcman 1985). Yet what is clear through the case study outlined above is the evolving social context which has brought about the use of forceps and the medicalisation of childbirth. What also becomes apparent is that regardless of the social intent for the design of the artefact such as the forceps, when used within our present medical structure, it becomes a political tool for maintaining the prevailing control over childbirth and hence women.

> JOURNAL ENTRY 1.10   *The social shaping of technology*
>
> Can you think of other examples within our recent history which demonstrate how the social context has evolved technologies that yield power?

APPROPRIATE TECHNOLOGY

TECHNOLOGY TRANSFER: THE SNOWMOBILE

Siraj-Blatchford (1997) provides an interesting example of how technology produced in one culture is inappropriately transferred to another.

continues…

> 'When they (snowmobiles) were purchased for the use in reindeer herding in Lapland, owners found they needed to carry ample supplies of fuel and spare parts, and to acquire new skills so as to accomplish emergency repairs when breakdowns occurred. The social impact of the snowmobile was, however, much more profound than this. The capital outlay and the expenses of maintenance meant that relatively few families were able to participate in herding by snowmobiles. Those who adopted the technology found it more economical to work with larger herds; as a result, small farmers, previously with their own herds, were bought out, becoming waged labourers or unemployed. The net effect was that a predominantly egalitarian society, where all owned and worked their own farms, was transformed into a duel society, inegalitarian and hierarchical. Indigenous industries associated with previous methods of herding, involving sledges, skis and dogs, were adversely affected by the change, and an increase in dependency on foreign sources of snowmobiles followed' (Layton 1992a, p. 42, cited in Siraj-Blatchford 1997, p. 29.)

The introduction of snowmobiles into Lapland provides an example of the mismatch between the values of the designer to the values of the consumer. The transfer of technologies from one culture to another are likely to yield long-term problems—as occurred with the Lap people who adopted the new technologies. Technology is constructed within a particular culture to meet the needs of that culture and therefore cannot be easily transferred.

JOURNAL ENTRY 1.11   *Technology transfer*
Can you think of other technologies that have been transferred from one culture to another? What have been the results?

Many aid programs operating in poor countries (often referred to as 'developing countries') have caused many long-term problems to the fabric of the community. Deliberate attempts are now made to develop technological solutions from within the context of the community in which the need arises—using local resources, local skills and incorporating local technologies. These technological solutions do not fit within the popular assumptions about technology, as Siraj-Blatchford reminds us:

> *People equate technology with 'advanced', 'sophisticated', 'high' technology, and these are value-laden words. But what kind of technology is really sophisticated? People don't equate the best technology with the most 'appropriate'; they equate it with the most complex or, even worse, the most expensive* (Siraj-Blatchford, 1997, p. 37).

A further consideration in the analysis of culture and technology is the exploitation of poor majority countries. There are many examples of natural resources (e.g., genetic material) which are scarce or no longer exist in rich minority countries (often referred to as 'developed' countries). This is often as a result of over-consumption and eventual depletion (plant, variety rights) land abuse and mass production in rich countries (e.g., of only a few varieties of wheat being taken from poor majority countries, patented and sold back for exorbitant prices).

Ozone-reduction standards which are currently being applied are also a case in point. The extensive development and over-consumption of resources in rich minority countries has meant that environmental problems have arisen. Standards being set for countries across the globe do not appear to take into account the rights of poor majority countries who also wish to enjoy technologies such as airconditioning and refrigeration.

A further example of exploitation is that of the Bhopal disaster.

> In December 1984, a gas leak at the Union Carbide chemical plant in Bhopal, India claimed over 2600 lives within a week. The majority of those who died were children. It has been estimated that the leak has led to the death of 10 000 people and as many as 300 000 have suffered serious damage from the methyl isocyanate that was leaked. The safety devices that the American company incorporated in its plant in Bhopal were inferior to those fitted in the United States, in France and in Germany. After a previous accident in 1982 the trade union at the plant printed 6000 posters warning of the impending danger and in October 1982 the workers launched a hunger strike to warn of the dangers. Nothing was done to avert the disaster.

Similar arguments can be directed towards rich minority countries such as Australia who purchase at very low prices goods from some countries. The low prices reflect the substandard conditions that young girls and women work in to produce goods for sale overseas. The unsafe working environment that many people work in should be foregrounded when countries make decisions about imports. Legislation which forbids the purchase of goods from companies that exploit their workers should be put into place.

> JOURNAL ENTRY 1.12 *Exploitation and technology*
> Can you think of other ways that Australians can change their practices so as not to endanger or encourage exploitation of poor majority countries? How can these issues be considered by young children? Siraj-Blatchford suggests:
>
> > *In choosing not to consider certain aspects of the subject we make political and moral decisions that influence our pupils' understanding of the subject, and of their role within it* (p. 30).

TEACHING TECHNOLOGY TO YOUNG CHILDREN

Technology education is a newly defined curriculum area in Australia. As a result, it is likely that you would have had limited experience as a technological learner in school.

> JOURNAL ENTRY 1.13 *What are your feelings about teaching technology?*
> Record what you think technology education should involve. How should this area be taught to children?
> Briefly consider what are the important areas that come under the umbrella of technology education.
> Should technology be taught to all children from birth to 12 years?

In broadening your views on what technology is, you may have also come to think more laterally about what technology education is. There are a number of approaches to the teaching of technology to young children (see Part 2 for details). The approaches adopted by teachers will reflect their beliefs about technology and technology education and the priorities set by departments of education.

SUMMARY

After reading this chapter you will be in a good position to revisit your views of what technology is (journal entry 1.2) and you will find that you are likely to add to your views. Continue to look at your views as you work your way through the chapters in this book. The intention is for you to develop your views, culminating in your own personal position on the teaching–learning process in technology education.

REFERENCES

Barnes, B. (1985) *About Science*, Basil Blackwell, Oxford.

Chambers, W. (1984) *On the Social Analysis of Science*, Deakin University Press, Geelong.

Davidson, N. & Rakusen, J. (1982) *Out of Our Hands*, Pan, London.

Ehrenreich, B. & English, D. (1973) *Witches, Midwives and Nurses: A History of Women Healers*, Feminist Press, New York.

Ehrenreich, B. & English, D. (1978) *For Her Own Good*, Anchor/Doubleday, New York.

Fleer, M. & Hardy, T. (1996) *Science for Children. Developing a Personal Approach to Teaching*, Prentice Hall, Sydney.

Forster, F. (1967) *Progress in Obstetrics and Gynaecology in Australia*, John Sands, Sydney.

Haire, A. (1973) 'The cultural warping of childbirth', *Journal of Tropical Paediatrics and Environmental Child Health*, June, pp. 172–91.

Hardy, T. (1997) 'Women's experiences of science and technology: diversity and tension in four case studies'. Paper presented at the 28th Annual Conference of the Australasian Science Education Research Association, University of South Australia, July 1997, pp. 1–27.

Jarvis, T. & Rennie, L. (1994) 'Children's perceptions about technology: an international comparison'. Paper presented at the Annual Meeting of the National Association for Research in Science Teaching Anaheim, March 1994.

Kitzinger, S. & Davis, J.A. (1978) *The Place of Birth*, Oxford University Press, Oxford.

Llewellyn-Jones, D. (1979) 'Pregnancy and childbirth', *Australian Family Physician*, 8(4), pp. 459–69.

MacKenzie, D. & Wajcman, J. (eds) (1985) *The Social Shaping of Technology*, Open University Press, Milton Keynes.

Oakley, A. (1976) 'Wise woman and medicine man', in *The Rights and Wrongs of Women*, J. Mitchel & A. Oakley (eds), Penguin, Harmondsworth.

Siraj-Blatchford, J. (1997) *Learning Technology, Science and Social Justice: An Integrated Approach for 3–13 year olds*, Education Now Publishing Co-operative, Nottingham.

Symington, D.J. (1997) 'Technology in the primary school curriculum: teacher ideas', *Research in Science and Technology Education*, 5(2), pp. 167–72.

Turnbull, D. (1991) *Technoscience Worlds*, Deakin University, Geelong.

Wajcman, J. (1991) *Feminism Confronts Technology*, Allen & Unwin, London.

Winner, L. (1985) 'Do artefacts have politics?' in *The Social Shaping of Technology*, D. MacKenzie & J. Wajcman (eds), Open University Press, Milton Keynes.

Chapter Two
———————

Technology in the home: children's experiences of technology

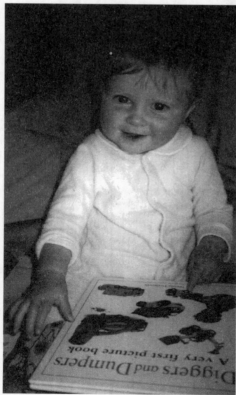

Figure 2.1 What technological experiences do children have at home during their school years and prior to beginning school?

INTRODUCTION

'I build things and play favourite games ... Knights. I play with my Lego and watch TV.' *(James)*

'These are new ... and these ... (points to everything) but my bedroom is usually messy. I play lots of games and I, we, do some work on the computer. I have a paintbrush and do reading and letters.' *(Lauren: entering a newly renovated bathroom)*

These two five-year-olds explain how they use their time whilst at home. What do we understand about children's home experiences? How will their technologically constructed environment influence how they play, interact, think and work at school, preschool or child care?

Children as young as five are now expected to be involved in technology education. Yet, we know very little about how young children should be involved in this systematically organised curriculum. Only a small amount is understood about the difficulties associated with introducing technology education to young children. The experiences children have in the home and community prior to coming to school influence their technological capabilities. For example, experience of high technologies such as computers, videos, cameras, TVs, and computerised toys will influence how they use equipment whilst at school or preschool. However, the experiences children have had with the technological process will also influence how they think and work. For example, children who have participated in drawing plans before making something, using templates such as sewing patterns, making lists in planning for a holiday or shopping trip, will bring to technology education valuable technological skills and capabilities. Culturally specific experiences such as drawing from a plan view in the sand or painting symbols in art work from a plan view will also influence their ability to conceptualise and work technologically (Fleer 1995). Children's prior experience with technological products and processes needs to be understood before commencing the planning process for children's learning.

JOURNAL ENTRY 2.1   *Children's technological experiences*

Record your responses to the following questions in your journal:

- Is it important to find out what children know and can do before planning for technology education?
- How will children's home experiences influence how they make sense of the technological learning we provide in classrooms and centres?
- What does it mean when children come from homes which use a range of high technologies (such as computers) or homes which do not have any? Does it matter?
- How can we find out what children know and can do before they begin school, preschool or child care?
- What cultural experiences do children have which are likely to facilitate engagement in technology education?

This chapter explores some of these questions and focuses predominantly upon children's experiences of technological processes. In particular, children's planning or designing will be considered. Children's experiences with making and appraising will also feature. Children's use of or exposure to more recently developed technological artefacts (e.g. CD Roms, Tamagotchi) is not taken up in this chapter. The emphasis on designing, making and appraising is taken up in this chapter because recent curriculum developments focus predominantly upon technological processes. A more detailed discussion of this approach to technology is considered in Chapter 6.

THE TECHNOLOGY LENS

> JOURNAL ENTRY 2.2   *Can children's play be thought of as technology?*
>
> 'Sometimes I run round and round and round the circle of trees and I play on my swing set. I make mulch and collect sticks to burn … collect firewood. I bounce a ball and I catch butterflies. Me, Daddy and Mummy work on the computer.' *(Jessica on tour of the garden)*
>
> 'I play with a helicopter … flying, and I play with my train set in the lounge room with Scott. Outside I jump on the trampoline and play in the cubby … but we can't play there at the moment, it's broken. We also play in the bungalow. In my room I like to read and play with teddies.' *(Regan)*
>
> 'This is my front verandah … I sit. I like to play with my barbie … do some paintings. I play on the quadro and draw on the chalkboard.' *(Erin)*
>
> 'Here is my trampoline … here are the budgies … we look after them and put food in the bowls. These are chooks … we give them food to eat (shows bag). These are my clothes I need to go to work, if robbers come.' *(Elliot on tour)*
>
> Should the play of Jessica, Regan, Erin and Elliot be considered as technology? In Chapter 1 the issue of what can be viewed as technology was considered.
> Can these children's experiences also be thought of as technological?
> Record your views in your diary.

Planning activities

If we examine the activities of an infant, toddler and preschooler in the home we begin to see how the child's culture involves them in a multitude of planning opportunities. Whilst each child's family experience will vary, the range of possible technological activity could be quite vast. Rituals and routines, whilst not always articulated to the child, do form an important part of the child's ability to predict or plan what will happen. For example, children from Western cultures are involved in dressing, shopping, cleaning the house, washing, bath time, singing games, peek-a-boo, and bed time, to name but a few important processes for the child. Similarly, unusual events such as going on an excursion or to a party are usually preceded by oral planning. Preparation for a visitor; using a manual to set up a video, tune a car, set up a sewing machine; using a plan to assemble furniture, follow a street directory, TV guide; and shopping centre guides are all technological activities that could take place in the home or surrounds. Similarly, traditionally-oriented Aboriginal children living in remote communities are likely to learn routes from one community to another through storytelling and art.

We need to know more about the key rituals that children engage in which form an important part of their daily planning and which provide a foundation for effectively engaging in technology education at school/preschool. In Tables 2.1 to 2.3, the results of an interview with a group of three- and four-year-old children who attend a child-care centre and responses from a group of preschool children is summarised. The interviews show from a Western child's perspective what technological tasks they are involved in when in the home.

Table 2.1   Routines

Child-care responses	**Preschool responses**
General routines	
Puzzles and draw, look at picture books and go out and ride my bike, and go down to the horse paddock. (Claire)	I have my lunch then I go outside and play. I jump with Tamara and Danielle. (Regan)
I sometimes get my backpack ready, before sleep time in the day I will probably watch Play School then go out in my garden and pick some lovely flowers. (Grace)	I have lunch then I play in my room. Then I play outside and chase butterflies and play with toys and make aeroplanes. (Jessica)
	I go into the pool. I have a sleep. I eat. (Erin)
I do drawing on my own. (Matthew)	I play with teddies and play with all my toys. Lunch is after I play, then I take Radar (the dog) down to the lake. (Elliot)
Play, a drawing. (Teddy)	
Watch the music box, watch *Blinky Bill* at Dad's place. (Sarah)	I have lunch and then go outside and play. I play with Robert. (James)
	I take off my shoes to lie down on the couch. Then I have my lunch and then afternoon tea. Then I have a big drink of water and then have another rest. (Lauren)
Morning routines	
Get dressed, and go in the car and drive, I do what my mummy and daddy says, I choose for playing. (Daniel)	I get some clothes on. I get dressed and put on my shoes and make my bed and get my bag and then I am ready to go to preschool. I look at the weather, it tells me if it is sunny or cold. If it's sunny I wear shorts and T-shirts, and if it is cold I wear a flannel shirt. (Regan)
Put your clothes on very quickly then go to the day-care centre. (Anthony)	
My mum would decide what I'm going to wear and, I decide what I'm going to play with. (Claire)	I have breakfast then I brush my teeth. I look in my cupboards and have a look at the clothes I want to wear. Now I look in my summer cupboard. (Jessica)
Mummy decides what I am going to wear. (Daniel)	I tell myself to get dressed. I wear what clothes my mum puts out for me. (Erin)

continues…

I don't know. (Anthony)	
I just think what I am going to wear? (Matthew)	I wear gloves, these are my motorbike gloves. Today I need a hat to keep the sun off my face. (Elliot)
I think those ones, we choose these and they got Bubby and Bubby shoes and Bubby pants and um, I have toys in my room and pokies. Mummy does (choose clothes). (Sarah)	I always get up and play with my Lego first. I wear clothes. I just know what to wear, I know when it is hot. (James)
	I choose what clothes to wear and Mummy puts them on. I look in my wardrobe and because it is cold I know … Mummy told me and I could see a grey sky. (Lauren)

Table 2.1 demonstrates a range of child-focused activities. In many instances the children have articulated these activities in the form of a progression. Most of the children have clearly expressed their ordered and planned world. With little prompting the children have been able to outline how their day is spent, with some making comment on how decisions are made with regard to these events. The act of planning is expanded upon in Table 2.2 where the children from the child-care centre outline what they understand about the word 'plan'.

Table 2.2 Planning activities (child-care centre children)

Level 1	Confusion	Planting. You're doing something. (Anthony)
Level 2	Emerging ideas	Yeah. Planting. Got to think of something. Don't know what it means. (Teddy)
		Um, I, I can plan, I can plan my train tracks and my, I can plan, playing with the train track and play which train I like. Um, I've got a book of planning about a cat. (Daniel)
Level 3	Observational	That means, I know what it means. It means doing hard work, hard work. My dad is a worker. Sometimes he does a bit of planning. He probably does a bit of planning of work. I think he just, the only, he talks to people on the phone and plans the … people that help him. (Grace)
Level 4	Event-focused	Planning to do some things. Going to someone's house. (Matthew)
Level 5	Construction oriented	When you plan something, you've got to, you can, you can plan something and then build it and, or what you want to do. (Alyse)

Five levels of thinking were evident in the responses given. Three children gave responses which demonstrated confusion (Anthony) or emerging understanding of the term (Teddy, Daniel). The term 'planning' was confused with 'planting'. However, two of the children were able to outline that it had something to do with *thinking*. Grace's understandings related to observing her father actively plan on the telephone. Similarly,

Matthew related the term to planning to visit someone. Alyse had a much broader understanding, she considered planning within the building process.

When planning was contextualised within a special event such as planning for a holiday or dinner, the following responses were given by the children (Table 2.3).

Table 2.3   Planning for special events

Child-care responses	**Preschool responses**
Holidays	
We need my nighties, or my pull-ups. If I've got enough. Some clothes. Some bedtime books. (Claire)	Mum decides. (Jessica)
We need bedspread, clothes and camera. (Daniel)	They ring up on the phone and we talk about how we are going to get there and see if we need a car. Then we ring up the person. We also need to think about clothes. (Erin)
I just, um, think. Well, some of my toys (that's all you will take?) Yeah, because they already have drinks at Grandma and Grandad's house. Or food, I don't need to take any food either. I just need to pack clothes and toys. (Matthew)	We go to the lake. We need to lock the house. We need to take food. (Elliot)
Beach ball, shovel and spades too. And a bucket. Food—meat you can eat, rolls, everything. Pillow and combs, toothbrush and tooth paste. (Teddy)	My dad decides. We pack our bags and go. He first sees if we all want to go. We need to lock the house. (James)
Need to take ... my water, it's all empty, have to get some milk instead, have to get some apple juice and um, some biscuits and um, bit of lunch and um, toys and oh, and I need um, my big toy, he's pink, Super Ted, and we'd need oh, little Ted and the Grandma Teddy Bear and the Poppy one. He's pink. Need some pencils, paper and colouring books, I think that's all now. (Prompt—clothes): Um, barbie, barbie, barbie shoes, barbie pants and one singlet. (Sarah)	We usually plan by thinking, we think what place we are going to and then we catch a plane. (Lauren)
Cooking	
Flour, pancakes. Flour, sugar, butter, mix it up and cook it. (Claire)	Mummy decides and tells me that I will have chicken. (Jessica)
Well, flour, cornflour, eggs, butter, margarine. We could make sprinkle cakes or you could make cream cakes... (Matthew)	Us ... we say what we want for tea. We have a meeting and discuss—only us (Tim and Erin) and then we tell Mum we want spaghetti. (Erin)
Ingredients. Chicken, potatoes, corn and that's all ... (Teddy)	Pizza. Geoffrey and I like pizza, Mum knows that. (Elliot)
Hot chockies with Grandma. You put milk ... you put the ... in the cup and you put, and there's chocolate in the top. (Sarah)	We have what we feel like. I just ask Mum to have what I want. (James)
	They choose it in their head and then they get an idea and then they use a cookbook to get the recipe. (Lauren)

Planning for these children is clearly something that is quite familiar to them. Their responses indicate portions of processes that they are likely to undertake. For example, Teddy speaks in categories—toys, food and then toiletries. Erin details how the planning process operates—phoning, travel requirements, and then packing. In the cooking example, all the child-care children detail the types of ingredients they are familiar with, each labelling what they are cooking. The preschool children discuss how they plan what they are going to eat. Although the focus for the child-care and the preschool children was different, their responses indicate an awareness of planning for cooking (ingredients, decisions regarding what to cook).

What is interesting to note in each of the three tables is that planning for the children is essentially oral. The children have not made references to writing things down. One would expect that there would be some two-dimensional planning occurring in these families, such as writing a menu or a shopping list. However, child involvement in the formulation of lists is likely to be limited—although requests from the child may be added. In some families, lists of things to be done may be drawn up. However, only oral planning (as opposed to written planning) was mentioned by the children in all interviews conducted with them, except for the following comments, which resulted from asking the children about going shopping.

'I tell Mummy what I want to buy. We have to write a shopping list.' (*Jessica*)

'We write a shopping list, we have to plan what we are going to buy.' (*Lauren*)

'We write down what we want on a list. But we first look in the cupboard and see if there is nothing.' (*Erin*)

There are three types of planning that are possible, oral, two dimensional and three dimensional. The least likely form of planning that children would participate in at home or observe family members engage in is three-dimensional planning or model-making. It is possible that in craft-oriented families some modelling may occur, however, it is likely that only the adult will engage in this activity and not the child. This form of planning was not mentioned by the children. Their responses provide an indication of the predominance of planning young Western children are likely to experience.

If young children's experiences prior to school involve mostly oral planning, with minimal two-dimensional planning and very little or no three-dimensional experience, it is little wonder that children do not intuitively engage in two dimensional or three dimensionalplanning/design work in school. Most of their planning experience is oral and hence children will use this mode for planning and designing. A great deal of experience with two dimensional (written/ drawing) and three dimensional modes for planning and designing would be needed by children if they are to engage in anything other than oral planning when in preschool, child care or school.

Making activities

Children participate in a range of activities in the home where they make things. How children come to understand the materials and equipment that they use is well documented. Infants have a great deal of experience with the oral exploration of materials. By the time

Figure 2.2 Children's home technological experiences will influence how they work at school, preschool and child care

children attend school they already understand a great deal about the properties of natural materials such as water, sand, air, rocks, leaves and bark, and processed materials such as metal (pots and pans), plastics (kitchen containers), glass, paper, cardboard and fabric. Yet their experiences with regard to adhering or joining materials, cutting materials, combining materials or changing materials to make something new are less well understood. Similarly, their experiences with different types of construction kits such as Lego will vary depending upon opportunity and adult intervention, interaction or modelling.

The sets of materials children are likely to experience in the home context include:

1. Recreational—jigsaws, craft work and model building.
2. Home environment maintenance—garden, house.
3. People-focused—food, baby care, sewing.

Table 2.4 shows the responses from a group of Western children to a series of questions on things they made with their family, or saw family members engage in (responses by child-care children are grouped together since responses were merged—Question 3).

Table 2.4 Making activities

Child-care responses	Preschool responses
1. What things do you do or make with Mummy?	
	We can't do much because she is always busy (after prompting). Yeah, I do help her make patty cakes. (Regan)
	Sometimes we sweep up the wisteria and we water the flowers and strawberries. We look after the pussy cat and hang the washing out. We do drawings, get the firewood and I help Mummy with flowers. (Jessica)
	We make cakes and do the washing. Outside we plant flowers and go for walks. (Erin)
	I help Mum plant, cook the dinner and make pictures and put frames around them. I draw with Mummy and go riding. (Elliot)
	I make cakes and things ... muffins, pikelets and pancakes. Outside Mum helps me build. (James)
	We make biscuits and cakes and lots of recipe things. We cook in the kitchen. I usually do some watering with Mum. (Lauren)
2. What things do you do or make with Daddy?	
	We build a cubby. (Regan)
	I collect firewood and help Daddy prune the apricot tree. I help him make dinner. (Jessica)

continues...

Child care responses	Preschool responses
	We swim and play in the water and we go to the shops. We draw pictures. (Erin)
	I help Daddy split the wood and make a big pile. We make books and we made a bird feeder. (Elliot)
	We make castles because he (dad) used to make castles out of rock. (James)
	I go to his work sometimes … I play on the whiteboard. Daddy uses it when he goes to meetings. (Lauren)

3. What does your mummy or daddy make or do at home?

Child care responses	Preschool responses
They make iceblocks, my Mum makes iceblocks when Mum gets home because I'll be able to have some spaghetti bolognaise. She makes teddy bear jumpers. (Alyse)	She does the washing and she sometimes makes porridge. She makes cakes and I help her make them, I put the butter in.
Some sewing. Mum does sewing. Dad doesn't know how to do them He knows how to make, to cook fish fingers. He makes lunches. (Claire)	He works in the nursery and sells plants at the markets. (Regan)
	She does the shopping.
	He makes the fire. (Jessica)
(After prompting) Making a book shelf Mm, … he got a piece of wood, put some glue onto them then got another one and stuck it down to the other, and stuck the two ones together stuck, um the middle one to the to the other end where the other ones are, ending and then, the um, moved another one onto the top and the bottom and then, and then, it was and then it was and then he just had to put another layer on top … and he had to varnish it. (Claire)	She makes cakes and food for catering and she makes my bed.
	He cleans the swimming pool. (Erin)
	She goes out, she goes horse riding She makes our beds and she looks after me. At work she mows gardens.
	He goes to work, he drives a truck. Daddy's motor bike is broken, it has a puncture so he can't ride it. He is trying to fix it, he has to take a screw off and put oil in it. (Elliott)
Um, no they don't. They only sit around and eat tea. (Grace)	I don't know! (long pause) She does the washing up and gardening. At work she looks after sick people.
(Mummy) makes cakes, play dough, everything like that. (Matthew)	He just works. He sometimes works on his trainer, it has wheels, they move but you can't ride anywhere. (James)
(Mummy) do some jobs. Around the walls she … painted. Make sandwiches … cuddles and kisses and, watches videos at the same time. (Sarah)	She usually cleans the house. She dresses me. She goes to her work and looks after people there.
	He usually is exhausted from his work and he lies down on the sofa and watches TV. (Lauren)

The children's responses to making things with their parents or observing their parents make things indicated that a great deal of making was occurring in each family. This finding is not unexpected. Most responses related to the maintenance of the home and family, with children participating in most events. Categories that emerged included: sewing, cooking, painting, lunch preparation, assembling of shelves, house cleaning, shopping, playing, washing, firewood collecting and gardening.

An analysis of the children's making activities in terms of materials, information and systems indicated that approximately two-thirds of all responses given related to making with materials (particularly food). The other third of responses indicated that children are involved in or observe activities that include systems such as routines and garden watering processes. Comments regarding designing, making and appraising with information technologies such as TV, letter writing, radio, computers, books or audio tapes were heard less frequently. These areas were considered by children when asked about their routines. However, when children were asked to comment on making activities, these areas rarely featured. It would seem that information technologies were more associated with passive viewing or receiving and not active designing, making, and appraising. For example, the children did not talk about constructing their own audio tapes (for stories, singing etc). Once again this is not surprising. However, it does highlight the need for curriculum developers and educators to be aware that the use of information technology needs to be reconsidered by children—from passive to active use.

Children's understanding and experiences of appraisal of processes and products were also sought during the interviews on making things. However, the children did not volunteer information regarding this area. Although appraisal is regarded as equally important to making and planning, the appraisal activities are much more difficult for very young children to discuss (metacognitively speaking).

Children's home experiences which featured technology tended to focus on people and home maintenance. For the children, most of their activities related to using materials. The children had less experience with information technologies. Appraisal comments were not forthcoming. What is interesting to note is the mismatch between children's experiences at home and what teachers are expected to plan for in their technology programs. Given the predominance of making and oral planning experiences of young children, as educators we need to pay more attention to planning experiences which will help children engage in two- and three-dimensional planning/designing (and possibly explicit discussion on appraisal). The experiences of children from Western cultures in this area are minimal and hence many free-play opportunities (and teaching modelling) of two- and three-dimensional planning/designing may be necessary if children are to feel successful in design, make and appraise with materials, information and systems.

The examples given, and hence the perspective inherent in this chapter, is that of very young children from Western cultures. Thought should be given to how the experiences will be different for children from different cultures. (See Chapter 3, 'Multiple world views in curriculum design and implementation: cultural constructions of technology'.)

SUMMARY

In this chapter young children's experiences in the home were deconstructed to determine what type of prior and concurrent technological activity they engage in. You were invited to think about these experiences in relation to your growing understanding of what technology is. Similarly, you were asked to consider how children's experiences will influence what they do at school, preschool or child care.

> JOURNAL ENTRY 2.3 *Expanding your views on what technology is*
> Record your views:
>
> - After reading this chapter, how would you define technology?
> - What technological experiences do very young children have prior to commencing school, preschool or child care?
> - How can you best build upon what children are capable of doing?
>
> Now compare your responses with your initial journal entry for this chapter.

REFERENCES

Fleer, M. (1995) 'Does cognition lead development, or does development lead cognition?', in Marilyn Fleer (ed.) *DAPcentrism: Challenging Developmentally Appropriate Practice*. Australian Early Childhood Association, Canberra, pp. 11–22.

ACKNOWLEDGMENTS

This chapter is a modified version of a paper published in the *Journal of Australian Research in Early Childhood Education*. Permission has been granted to reprint the research paper in this chapter.

Fleer, M. (1996) 'Investigating young children's home technological language and experience', *Journal of Australian Research in Early Childhood Education*, vol. 1, pp. 29–46.

Special acknowledgment to Marita Corra and Wendy Newman is made. Their contributions to the research are much appreciated.

Chapter Three

Multiple world views in curriculum design and implementation: cultural construction of technology

INTRODUCTION

As mentioned in the section 'The social shaping of technology' in Chapter 1, throughout history technologies have been developed in social contexts.

> 'Mainstream schooling is only one aspect of an Aboriginal child's learning experience. Others are learning about the land, role playing of animals, fishing and the laws and customs of the culture.' *(Donna)*
>
> 'Aboriginal people have developed many useful tools/implements and weapons that were effective in doing their intended tasks. Aboriginal people incorporate learning into everyday life. My own experience is that when I was young my uncles and grandfather would go camping and do a bit of fishing. They would show me how to do it and talk about what the tool is used for. This type of scenario happened all the time, the elders always felt the need, and thought that it was a necessity, to speak of the certain characteristics of everyday tools and plants and their usage.' *(Jonathon)*
>
> 'In traditional Aboriginal lifestyles, the use of technology was a part of daily life. For example, in hunting many different devices were used to produce the most effective method of collecting food. For hunting big game the woomera was used as a projectile sending the spear much further than if thrown alone. The boomerang was used as a hunting tool. Differently shaped boomerangs served different purposes. Stones were used widely for making knives, axes and spears. Woods and grasses were used for protection (e.g. a shelter) and making fires.' *(Judy)*

These Aboriginal pre-service teachers were sharing their ideas about learning and technology from an Aboriginal perspective. What would technology curricula look like if it included and respected Aboriginal culture? We see one answer to this question in the following vignette.

TECHNOLOGY CURRICULA FROM AN ABORIGINAL PERSPECTIVE

If we look at curricula written from an Aboriginal perspective by Jane Proctor we notice that it is holistic, and incorporates aspects from both the past and present.

In a technology sequence for primary school children Jane decided to focus on the use and making of the spear as an instrument. She began the first lesson by considering the history of the spear, and how it was traditionally made by fire-burning, and bending of blackwood into the spear shape. Special rocks were used to shape the head of the spear by chipping the rock into a triangular-shaped spearhead. Special wood was used to make the spear handle, and the spearhead was fastened to it using kangaroo skin shaped into string. The spear was mainly used as a hunting instrument, but it was also used as a digging stick, as a tool for knocking fruit from high trees, and fishing for eels and mussels.

Jane showed the class examples of traditional spears, as well as spears that her husband had made, which he only used for fishing purposes. The consequent class discussion focused on how he made his own spears using modern equipment, and materials such as steel, iron and wire. He told the children how he looks for a very strong steel rod from which to make the spear handle. Then he gathers scrap metal and looks for smaller, thin rods to be sharpened by filing the ends and making sharp points. He attaches thin rods with his wire to the handle which is about the size of a broom handle. He then goes out on hot summer nights, spearing for fish and eels in the local rivers. The children were then asked to design their own spears using raw materials such as sticks, branches, vines and rocks.

In Australia the Aboriginal people have been practising science and technology for more than 40 000 years, long before white settlement 200 years ago. They use their knowledge about the movement or migration of animals for hunting. Their excellent understanding of the seasons and weather changes means that they know what fruit is ripe at what time, and when there is an opening of the lakes (mouth) everybody goes fishing or spearing. In addition to the spear, they design instruments for survival which include the boomerang, woomera, digging stick, handmade baskets for carrying, mixing paints for artwork, making of clothes, making of fire, identifying areas to live in and making camps. Each instrument has its purpose and the technological knowledge was important because if the instruments were not designed and made properly they would not work. The Aboriginal knowledge is shared and passed down over generations as life skills, by speaking, drawings and dance.

(Adapted from Jane Proctor's notes)

Jane's use of the strategy of including her husband's method of spear making is a good way of making the curriculum more culturally inclusive. If at all possible, teachers should invite Aboriginal people to come along to school and talk to the children so that they can hear indigenous people's explanations. This approach is consistent with the view expressed in Chapter 1 in the section 'Technological history'. We cautioned non-indigenous groups about commenting on indigenous technology when they do not have all the cultural information needed to make informed decisions. The point was made quite strongly against curriculum materials which encourage European children to evaluate technologies from cultures other than their own.

> JOURNAL ENTRY 3.1   *Cultural values and technology*
>
> In outback Australia Aboriginal people designed spears for survival purposes in a harsh environment. We included this example to focus your thinking on how you might design a technology curriculum which is culturally inclusive.
>
> How do you feel about encouraging children to investigate, design and make spears?
>
> Would you be comfortable including this activity in your curriculum? Why or why not?
>
> When answering these questions did you make judgments based on your cultural values?
>
> What other forms of technology did the traditional Aboriginal societies use to satisfy their basic needs?
>
> What may be appropriate in one culture may not necessarily be appropriate in a different culture.
>
> Start writing a list of indigenous technology which children could design and make that would be acceptable to both indigenous and non-indigenous people. As you progress through this book make additions to your list of culturally inclusive technology.

At this stage you should be aware that technology has intellectual, practical and ethical dimensions. A particular design solution appeals to a set of values, and any decision made when designing is value-laden. Every decision regarding the choice of materials (What effect does the material have on the environment?) or making an improvement (Who is the improvement benefiting?) involves an ethical choice. When children engage in technological activities they learn to stand by their decisions as they commit themselves to turning their designs into reality. We know that these decisions will be influenced by the children's cultural background.

In Chapter 1 you were challenged to think about how technology can be defined. You probably found that this task was not as straightforward as you initially thought. This situation is not unexpected, given that: 'Technology is a multifaceted entity. It includes activities as well as a body of knowledge, structures as well as the act of structuring' (Franklin 1992, p. 14).

Throughout the present chapter you spend time considering technology as practice, because it shows you the deep cultural link of technology. Ursula Franklin prefers to think of technology not in terms of systems but as a web of interaction. She reminds us that: 'Technologies are developed and used within a particular social, economic, and political context', and that to discuss the web of technology 'requires an examination of the features of the current pattern and an understanding of the origins and the purpose of the present design' (pp. 57–8). For this reason we now look briefly at the different approaches technology education has taken in Western countries.

TECHNOLOGY FROM A WESTERN PERSPECTIVE

Internationally technology in the curriculum has taken various forms in different countries, depending on the circumstances that led to its introduction. In the context of Western Europe, de Vries (1994) classified the diverse approaches to technology curricula as the Craft-Based approach, Technology Concepts approach, Design approach, Occupational or Vocational approach, Science/Technology/Society (STS) approach, High-Tech approach, Integrated Subjects approach and Applied Science approach.

JOURNAL ENTRY 3.2   *Diverse approaches to technology curricula*

Were you surprised by the number of different approaches to technology curricula?
   Name two approaches which you think are commonly practised in Australia.
   Which approach is predominant in your state or territory curriculum document for technology?

An example of the Design approach is Design and Technology (D&T) in the national curriculum in England and Wales, which emphasises design and capability. It is acknowledged that curriculum developments in the United Kingdom have largely influenced the direction technology education has taken here in Australia. Educators from Asian countries (e.g. Taiwan) are also turning to the D&T curriculum and comparing it with their own art and craft programs in an attempt to incorporate technology education in their schools (Tseng & Fang 1996). It is of interest to note that the Taiwanese government is spending 15% of their Gross National Product (GNP) on education (compared with 5% in the United Kingdom). Whilst currently its work is fairly conformist, it is rapidly progressing in the area of technology education by promoting 'living technology' (Kimbell 1997).

In Australia many teachers implement a design approach to technology by setting a challenge for their class in the form of a design brief for a group activity. Design briefs are the outline of the activity and tell the children the information they need to complete the activity. Figure 3.1 is a general description of the components of a design brief which you can refer to when you write your own.

TITLE

Introduction
Introduces children to the possible problem and asks them to help solve the problem.

Brief
Sets out what the children have to do. It explains what they are to design and construct.

Specifications
Sets out the constraints placed upon the children in respect to their design. Specifications also list the materials that may be used and special features that must be included.

Presentation
Explains to children how they must present their design and associated material when it is completed.

Time
Suggests the possible time allotment for the activity.

(Adapted from Country Education Project (1993))

Figure 3.1   Design brief

JOURNAL ENTRY 3.3   *Writing a design brief relating to survival in the bush*

We all need water to survive. Imagine that you were on an overnight hike and you accidentally lost your water bottle. Discuss with a partner how you might collect water in a bush environment. Use the design brief format above to write the criteria for the technological activity of designing and making a device to collect water in the bush.

If possible give this design brief to someone who has a different cultural background to you and ask for critical feedback. Make your design brief as culturally inclusive as possible.

Test out your modified design brief when you next go bushwalking, or set the challenge for children when they are on a school camp. Remember to pack a camera and take photographs of the solutions to the challenge.

In Chapter 6 design is considered in detail. In this chapter we look more closely at the cultural aspects of technology, and consider how we might incorporate Western, Eastern and Indigenous perspectives in technology curricula.

SCIENCE AND TECHNOLOGY FROM A MAORI PERSPECTIVE

Maori food and tools formed one of the topics included in a recent television program of *Gardening Australia*. The historical origins of the Maori were traced back to AD 925, when people from the Polynesian islands settled in New Zealand, bringing with them a variety of food and tools, and modifying their gardening techniques to suit the New Zealand climate. Examples of traditional tools dating back many hundreds of years have been found in swamps throughout the country. They include:

- **Digging sticks**, which are six to seven feet long, and are plain-shaped sticks, each with a point and a foot rest.
- **Spades**, which are similar to a European spade, only made out of wood.
- **Weeders**, which are v-shaped sticks and are used to extract weeds.

(Campbell 1998)

> JOURNAL ENTRY 3.4 *Maori food and tools*
> Maori are indigenous peoples of Aotearoa, New Zealand. How might you incorporate Maori food and tools in a technology curriculum?

Curriculum writers in New Zealand have thought about this question during the writing of national curriculum documents in Maori which address Maori language, science, mathematics and the social sciences. This curriculum reform began in 1992 and Maori education is now legitimate. The writing group for the science document was all Maori and consisted of seven trained primary teachers, six trained secondary teachers and one speaker fluent in Maori, employed in schools to support Maori language programs (McKinley 1996). As the writing group wanted the document to be firmly grounded in the Maori world with Maori values, it begins with a tauparapara. Tauparapara is a belief that the Maori world is made up of both spiritual and physical attributes.

During the writing of the science document there were many debates about the nature and ownership of knowledge and how 'Maori science' might differ from 'Western science'. Elizabeth McKinley, one of the project co-ordinators, contends that: 'knowledge in the curriculum needs to be relevant for a wide range of people in different situations at different times' (McKinley 1996, p. 165).

Think about the Australian situation. Could a technology document be written in a language of the Aboriginal and Torres Strait Island people given that there are more than 20 Aboriginal languages in use for day-to-day communication?

INDIGENOUS PEOPLE'S WORLD VIEW IN AUSTRALIAN CURRICULA

The issue of culturally inclusive curricula has been addressed in the Northern Territory with the recent release of the draft science curriculum, which incorporates an indigenous people's world view (McGinn, 1991, p. 130) and clarifies the meaning of a world view by defining it as 'a descriptive-interpretive mental model of the universe and its phenomena'. The science curriculum document is significant because:

> *In acknowledgment of the Territory's population of about 30% indigenous (Aboriginal) students, this is probably the first system-level science curriculum to recognise the value of the different worldviews of indigenous peoples, and it goes beyond the national statement which calls for inclusion of the science of various cultural traditions to inform Western science* (Michie 1997, p. 1).

Michie drew on his own experience of teaching in a cross-cultural setting, and the literature written on border crossing and multiple world views by Aikenhead (1996) and Pomeroy (1994). He recognises the overlap between the Western and indigenous world views, and urges teachers to acknowledge the particular world view they are operating from, so that both views can be valued and appreciated.

In Canada Aikenhead (1997) argues for facilitating border crossing when First Nation students study Western science and technology, because students from an indigenous 'traditional' background experience problems learning subjects grounded in Western culture. However, teaching from a static multicultural view often results in indigenous students being assimilated by its dominant Western culture. Pomeroy (1994) calls for science and technology educators to consider alternatives to assimilation by teaching for cultural diversity, which requires moving to a dynamic cross-cultural perspective through the following agenda:

- study the science in 'folk knowledge' or 'native technologies';
- bridge the world view of students and that of Western science;
- explore the beliefs, methods, criteria for validity and systems of relationship upon which other cultures' knowledge of the natural world is built.

To assist teachers to change their practice Michie sees a need for curriculum support materials which are designed to facilitate border crossing by making border crossing explicit for students, and which validate the students' personally and culturally constructed ways of knowing.

MULTIPLE WORLD VIEWS BASED WITHIN THE RESPECTIVE CULTURES

In order for curricula to be designed to accommodate these differing world views we must better understand the holistic way indigenous people view the world. This understanding can be gained by communicating with Aboriginal people, particularly the elders in the community. An alternative is to refer to stories told by Aboriginal elders, such as *A Day in the Bush*, written by Lucy Briggs and illustrated by Patricia Clarke. In this story an indigenous family spends a day in the bush where they use the following traditional technologies and techniques:

- a **digging stick** to find yams and grubs;
- a **dilly bag** to carry food;

- **tapping sticks** for use at rabbit holes;
- cooking underground using **fire stones**;
- a **mia mia** shelter to protect the family from the sun; and
- burying the scraps from the meal.

This story could be the focus when considering technology and sustainability issues, and the indigenous family's behaviour could be compared with Western people's practices. Discussion could contrast the tapping sticks which were used to stir rabbits from their burrows, with the Western approach of using ferrets. Cooking underground using fire stones can be regarded as an alternative energy source rather than using gas as a fuel. The practice of burying scraps is a form of composting and is one method of recycling.

Construction technology allowed Aboriginal people to dwell comfortably in all seasons. They needed to know how to build shelters using available natural materials to cope with severe heat and cold, winds and rain. A mia mia is a temporary home for Aboriginal people which is made from branches and leaves. Children at Hartwell Primary School were challenged to build a mia mia shelter as a technological activity in an integrated unit relating to Aborigines (Briggs-Pattison & Harvey 1998). The teacher explained that prior to this activity the children had been on a school camp where an Aboriginal elder told them about his life as an Aborigine and aspects of his culture. The teacher describes below how she introduced the activity to the Year 4 class.

> 'We were doing housing at the time. We sat down and talked about shelters, different types of shelters and what they know about Aboriginal shelters. We talked about the environmental effects and then went out into the environment. We talked about traditional shelter construction. In particular we discussed the context in which a mia mia would be made in the past and today. The children were given two lessons to make the mia mia.' *(Lynne)*

If you look at the photographs in Figure 3.2 of the children's attempts at making a mia mia after one lesson, you see that it is not an easy task.

When the children were asked about constructing their shelters they had the following to say.

> 'It was hard because it wouldn't stay up and kept collapsing.' *(Craig)*

> 'We were meant to overlap the thick sticks but sometimes they wouldn't reach, and then you had to go and find more sticks.' *(Fiona)*

> 'Whenever one stick fell it would knock another.' *(Jarrod)*

> 'It was important that the rain didn't come through or if it was windy that it wouldn't blow over.' *(Fiona)*

We can appreciate this example of technology more when we consider the factors that the Aboriginal people took into account when building a mia mia, such as: the use of leaves to deflect the rain, the placement of the mia mia in terms of its orientation to the sun and wind, the shape which allows for the warmth from the fire but does not allow the smoke to get in, and for storytelling when all the family members are situated closely together.

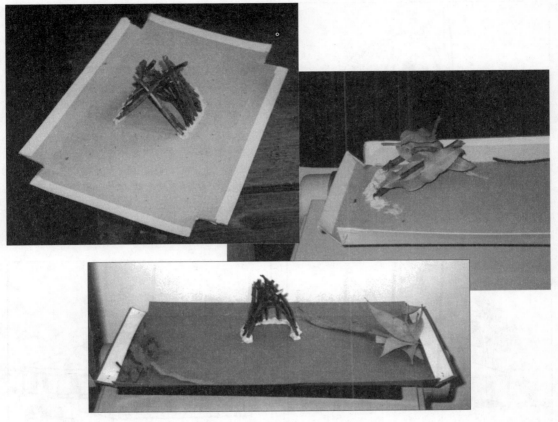

Figure 3.2   Year 4 children's initial attempts at making model mia mia shelters

> JOURNAL ENTRY 3.5   *Building a mia mia*
>
> How did traditional Aboriginal people live?
>   How do Aboriginal people live today?
>   Why might Aboriginal people want a temporary home?
>   Make a model mia mia using stiff cardboard for the base, plasticine rolled into a horseshoe shape, sturdy twigs for a strong framework and leaves.

'Aborigines varied the design of their houses according to the climate, availability of materials for construction and the likely period of occupancy' (Australian InFo International 1989, pp. 20–1). Living was predominantly done in the open and even today Aboriginal people prefer to live outside their homes as much as possible. You can see from the illustrations shown in Figures 3.3–3.6 the influence of Western technology on the style of homes, which in turn influences cultural practices. Particular homes were in many cases designed without consultation with Aboriginal people. The designers did not pick up on the cultural needs, or that the houses were not oriented correctly in relation to the sun and wind for cooling.

46 TECHNOLOGY FOR CHILDREN

Figure 3.3   Traditional housing

Figure 3.4   Stringybark hut on stilts

Figure 3.5   Outstation housing

Figure 3.6   Examples of housing in parts of the Central Desert

The media and technology

Newspaper articles make a good starting point for technological activity. The following article illustrates the misunderstandings which can occur when cultural values are not adequately taken into account.

CURSE FOR HUMPY VANDALS

Revellers who destroyed a symbolic 'healing humpy' at the Aboriginal Tent Embassy during the Canberra Festival could face an ancient curse, elders warned yesterday.

The party-goers knocked over the sticks-and-leaves humpy—one of several near the tent embassy in front of Old Parliament House—during the festival's fireworks show on Sunday night.

The embassy, which has been derided by locals as an eyesore, has been the centre of controversy for flouting a total fire ban by keeping a 'sacred fire' ablaze.

Elder Kevin 'Uncle' Buzzacott described the people who destroyed the humpy as 'corrupt' and said they would be cursed unless they apologised.

'If anything happens to them that's their bad luck—they will be cursed,' he said.

'This is a very special and powerful place. There is a lot of power in the humpy.

'They need to come to us and apologise, say they are sorry for the damage they have caused.'

Mr Buzzacott said the humpy faced Lake Burley Griffin to welcome people to the embassy and encourage healing. Mr Buzzacott said he had received assurances from festival organisers that they would fence the area off.

'They didn't do any of that—they had car parking so that people had to walk through our sacred site to get to the festival,' Mr Buzzacott said.

By Matthew Horan

Source: 'Curse for humpy vandals', *Herald Sun*, 11 March 1998, p. 27.

JOURNAL ENTRY   3.6 Healing humpy

In a small group in your tutorial discuss the ideas presented in the article.

How might you use the article as a basis for introducing a technological activity related to structures to children?

What technological task would you set for them?

Lynne had several ideas about how she might use the article with her class.

'I could give the children a photograph of the healing humpy without the text, and have them design a headline without knowledge of the text.

The children could extend the extremities of the photograph at either end and predict what would be there.

Linking with language, give the children the photograph and the headline and have them guess what the article is about and then retell it.' *(Lynne)*

The article above further reminds us that technology does not occur in isolation from values, beliefs and social life, and that a better understanding of indigenous technology will be gained by teaching the social and cultural mores associated with its use. For example, Aboriginal women generally do not touch didgeridoos, so they do not share in the technology surrounding the production and use of these instruments. On the other hand, Aboriginal women have women's business, and are more involved in the making of clothing, dilly bags and carry baskets. Some knowledge is shared, such as cooking, tanning hides etc. Marjorie wrote about the making of the didgeridoo in the following way.

> Aboriginal people use sound through the didgeridoo for many purposes e.g. communication, relationships, identities and at social gatherings. The sound of the didgeridoo is used for a variety of celebrations and ceremonies such as the corroboree. There is chanting from the women while the men play the sounds (representing many animals) from the didgeridoo. When particular sounds are played the Aboriginal people start dancing to the actions of that animal.
>
> To make a didgeridoo the men would go out into the bush and look for a particular type of tree branch. If the suitable branch was dry they would rub goanna oil all over it and then put it on an ants' nest for a couple of weeks. The ants would hollow out the branch. If the branch wasn't the shape they wanted, they would use a certain sharp stone to peel away and shape it the way they wanted it. The men would then use the ochre gathered from rocks and certain coloured bushes. They would use sticks that were previously placed in the fire to get hot, to burn designs (usually of their animal totems) into the didgeridoo. The didgeridoo was designed by the traditional Aboriginal people of Australia and it has a unique sound.
>
> (Adapted from Marjorie's notes)

Progress in medicine, weaponry and food preparation is a combined effort in Aboriginal communities. Small advancements are made continually so that ever increasing efficiency occurs. Indigenous technologies can be quite complex, such as the sophisticated large-scale networks of fish traps and canal systems for catching eels. Another example is the 'extractive technology' which enables the Aboriginal people to use highly toxic flora for food. Many early explorers in Australia suffered the effects of poisoning from eating the nuts of the cycad tree. Cycads are found in dry, open woodlands in northwest Australia in places like Cape York and Arnhem Land. Each plant produces 20–30 large nuts which, when ripe, hang beneath palm-like branches. The nuts are highly toxic in their unprocessed state because they contain the toxin 'macrozamin' or 'cycasin'. 'It is quite remarkable that Aborigines have developed the processing and cooking technology necessary to render the seeds of cycads edible' (Isaacs 1987, p. 82). These nuts are easily harvested by women who gather, crack and pound the seeds as a communal activity. The kernels are then leached (soaked for a considerable period in still or running water) and fermented into nut meal. This paste is made into damper and cooked in the ashes.

THE DREAMING AS A STARTING POINT FOR TEACHING TECHNOLOGY

Donna, an Aboriginal pre-service teacher, describes how she read a Birrirrh Dreaming story to a group of Year 3 children (three girls and three boys) who were non-Aboriginal.

> We reflected on how the Dreaming created the land, rivers, food, animals, tools etc.— that it did not mean sleeping. Most children told of stories, showing their awareness, rather than asking all questions. Some of the stories told by the children included

technological and scientific information. For example how termites are used to make the hole in didgeridoos. The children were told this in Year 2. Another girl told a story about sap on trees: 'that some sap can be eaten' and I said 'sap can also be used for glue'. This was a good lesson. Not only did we cover science, but we touched on racism and the children developed the concepts of how things are related to each other and discovered a different way of looking at creation through the Dreaming.

(Adapted from Donna's notes)

Dreaming stories are accepted as an accurate and valid history of Aboriginal people. 'Dreaming stories are designed to teach children about the spiritual world, the natural environment and rules for living' (Briggs-Pattison & Harvey 1998). With her group of children Jackie told a Dreaming story about the *Djang'kawu Sisters* and how they created the land from Burralku to Yalangbara. She contrasts this story with the Western view of creation.

The *Djang'kawu Sisters* also created the people, plants, animals, birds, waterholes and hills as they travelled west. This story is an example of technology, the specific scientific skills of navigation and geography. The sisters followed the sun and the morning star in a canoe until they landed on the beach. As they stepped out of their canoe it turned into stone and it still stands there today. So on foot the sisters continued to roam, creating the land and stopping along the way giving birth at different places. The sisters gave the people different sacred dillybags with their own design and different *likan* (totemic groups) of each *Dhuwa* clan. This Dreaming story is kept strong through our songs, dances, knowledge and culture. In Western culture it is believed that evolution of the apes was how people are like they are today and in Yolngu (Aboriginal people of eastern Arnhem Land) culture the *Djang'kawu Sisters* were responsible for life.

(Adapted from Jackie's notes)

The Dreaming is also the starting point in the technology units written by teachers at Clare Primary School, a country town situated 130 km north of Adelaide. The school is designated as an Aboriginal Studies Focus School with a Focus School Resource teacher. Their teaching programs featured the Dreaming. The units that the teachers developed started with the technological information contained in the Dreaming stories. Children reflected on these stories (prior knowledge) and realised that not all technology is new.

Take for example one unit called 'String' which the children began after completing a science unit on fibres and materials. After viewing a video of the Ngarrindjeri Dreaming story of *Thukeri the Bony Bream* told by Leila Rankine, the class discussed the aspects of technology in the story, such as making fishing lines, hooks, cooking technology and canoe technology. The technology unit involved the children in looking at string and string making, rope and rope making, nets and net making, weaving and toys. This unit fits nicely within the Systems strand of the nationally developed *A Statement on Technology for Australian Schools* (Australian Education Council 1994) and is suitable for Bands A and B (see below).

> **SYSTEMS**
>
> Make, assemble, organise, manage and modify systems.
>
> Students use natural materials to plan and construct manufactures which were and are used by Aboriginal people (e.g. fibres, reeds, sinew to weave into fabric/string for nets, mats and baskets; explore other utilisable resources).
>
> (South Australian Department of Education 1995, p. 11–9)

The teachers invited Laura Agius, a member of the community, to demonstrate her exemplary skills of Ngarrindjeri weaving. Laura explained how she valued being taught by her elders and how she in turn had taught her own children. She was pleased to teach others whom she knew valued the cultural information, language and skills she possessed. The intricate detailed fibre crafts of contemporary Aboriginal societies demands remarkable skills which have been handed down through generations.

EASTERN PERSPECTIVE: SHARING CULTURE THROUGH STITCHING

Textiles—the technology that cuts across cultural barriers

Textiles can often be a neglected area in primary schools and early childhood centres. Yet working with textiles can develop skills in sewing, weaving, knitting and screen printing, as well as knowledge about fabrics, fibres, dyes and clothing in different periods and cultures. Textiles is an area in which cross-cultural interactions are being fostered in England. In Birmingham members of the Aston Hall Asian Women's Textile Group are using embroidered textiles as a vehicle for cultural integration. The women who meet together to design and make wall hangings are now going into schools and working with teachers and children in their technology programs. These craft workers provide the children with experiences of learning stitching to produce embroidered objects. In this way quality products are being achieved by the children.

Working with artists and craftspeople (such as the Aston Hall Women's Textile Group) on specific projects provides valuable opportunities to develop the hidden curriculum through social and cultural interactions (Viegas & Benbow 1997). The photographs in Figure 3.7 show Indian women teaching workshop participants (technology educators) to sew mirrors on to fabric. Participants coloured pieces of cloth by tie dyeing, and then sewed beads and mirrors to decorate the fabric to make bags etc.

Figure 3.7 Indian women sharing ideas through stitching

With our multicultural society, this type of technological activity could open up many opportunities in Australia.

SUMMARY

In this chapter we have highlighted the importance of designing technology curricula which incorporate multiple world views. When implementing such curricula teachers should indicate which world view they are operating from, and facilitate border crossing for children. Teaching technology by starting with Dreaming stories is consistent with a constructivist view of learning, and this strategy values and respects an Aboriginal people's perspective. We have referred to the interwoven nature of culture, technology and science. The relationship between science and technology is deconstructed in the next chapter.

> *Culture, science, and technology, although distinct on specific levels, have been and continue to be inextricably bound to one another in such a fashion that each actually merges into the other, laying lines of contact and support. These relations involve a kind of complexity which prohibits us from claiming that any one of the three is distinctly prior, primary, or fundamental to one of the others* (Menser & Aronwitz 1996, p. 7).

REFERENCES

Aikenhead, G.S. (1996) 'Science education: border crossing into the subculture of science', *Studies in Science Education*, 27, pp. 1–52.

Aikenhead, G.S. (1997) 'Towards a first nations cross-cultural science and technology curriculum', *Science Education*, 81 (2), pp. 217–38.

Australian Education Council (1994) *A Statement on Technology for Australian Schools*, Curriculum Corporation, Carlton, Victoria.

Australian InFo International (1989) *Australian Aboriginal Culture*, Australian Government Publishing Service, Canberra.

Briggs, L. (1997) *A Day in the Bush*, Deakin University Press, Geelong.

Briggs-Pattison, S. & Harvey, B. (1998) *The Dreaming*. Scholastic, Hyde Park Press, Adelaide.

Campbell, C. (1998) Maori food fact sheet, *Gardening Australia*, 23 January 1998, Hobart.

de Vries, M. (1994) 'Technology education in Western Europe', in *Innovations in Science and Technology Education*, D. Layton (ed.), UNESCO, France, vol. V, pp. 31–4.

Franklin, U. (1992) *The Real World of Technology*, CBC Massey Lectures Series, Anansi Press, West Concord, Ontario.

Isaacs, J. (1987) *Bush Food. Aboriginal Food and Herbal Medicine*, Lansdowne, Sydney.

Kimbell, R. (1997) Technology Education—a thinking discipline? Presentation at Scotch College, 14 May 1997.

McGinn, R.E. (1991) *Science, Technology and Society*, Prentice Hall, Englewood Cliffs, New Jersey.

McKinley, E. (1996) 'Towards an indigenous science curriculum', *Research in Science Education*, 26 (2), pp. 155–67.

Menser, M. & Aronwitz, S. (1996) 'On cultural studies, science and technology', in *Technosicence and Cyberculture*, S. Aronowitz, B. Martinsons & M. Menser with J. Rich (eds), Routledge, New York, pp. 7–30.

Michie, M. (1997) 'Crossing borders: understanding differing worldviews of science through the Northern Territory science curriculum'. Paper presented at the GSAT and IOSTE Regional Conference, Perth, December.

Pomeroy, D. (1994) 'Science education and cultural diversity: mapping the field', *Studies in Science Education*, 24, pp. 49–73.

South Australia Department of Education (1995) *Aboriginal Perspectives Across the Curriculum*, South Australia Department of Education.

Tseng Kou-Hung & Fang Rong-Jyue (1996) 'A comparison of elementary school technology education in Taiwan and the United Kingdom', in Proceedings of the International Primary Design and Technology Conference, Birmingham, vol. 2, pp. 39–42.

Viegas, E. & Benbow, E. (1997) 'Stitching together', in Proceedings of the International Primary Design and Technology Conference, Birmingham, vol. 1, pp. 42–5.

ACKNOWLEDGMENTS

The vignettes included in this chapter were generated by Aboriginal pre-service teachers studying at the Koorie Institute of Education, Deakin University, Geelong. Thank you to Hartwell Primary School for the photographs and Lynn Dossetor for her ideas related to the mia mia shelter activity. Thank you to Maria Mason for her drawings on page 46 of this chapter.

Part 2
Approaches to teaching technology education

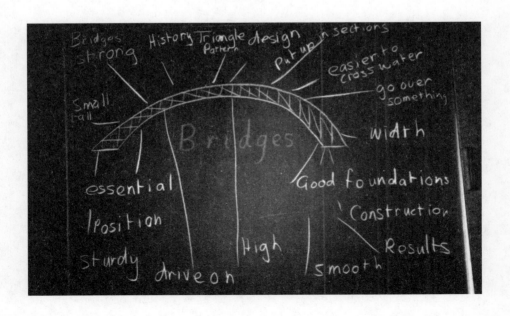

Chapter Four

Science–technology relationship

INTRODUCTION

> JOURNAL ENTRY 4.1   *Raising questions about ice-cream*
> Obtain a bowl of ice-cream and think about what ice-cream really is.
> Make a list of questions which come to mind as you observe the ice-cream in the bowl, as you taste the ice-cream, and after eating the ice-cream.

As part of a unit on dairy products Merilyn Costa, a teacher of year 6 students, encouraged her class to explore ice-cream using all their senses as you have just done.
Are your questions similar to their questions? Their questions were:

- How come it's soft even though it's 'ice'-cream?
- Why does it go frothy when it starts to melt?
- How much sugar does it have?
- Why doesn't it taste anything like cream?
- What are the ingredients?
- How do they make it?
- Is ice-cream just frozen cream?

Merilyn had planned an excursion to an ice-cream factory where the children could discover answers to many of their questions. At the factory they formed small groups and were taken to observe:

- cone making
- mixing of ingredients
- the testing room
- ice-cream and icypole making

continues...

- the freezers and machines that aerate the ice-cream
- the tubs of discarded ice-cream ready for remaking
- packaging.

On returning to school, Merilyn talked about the various processes and terms they had come across, like 'pasteurisation' and 'homogenisation'. The children had anticipated the production line workers, but had not thought about the people responsible for testing etc. The children were able to prepare a flowchart showing the stages of manufacture of ice-cream. They were given a list of ingredients to bring from home for the next lesson—ice-cream making!

MAKING ICE-CREAM

Ice-cream is basically made up of milk, cream and sugar. Commercial ice-cream has non-fat milk solids and modifying agents added and uses fresh milk and cream so it is necessary to pasteurise and homogenise it. Whipping air into the ice-cream mix gives it its smoothness, as does its rapid hardening. The flavour and texture are determined by the proportions of ingredients. Adding salt to the ice water reduces the temperature of the mixture rapidly. Salt has the effect of lowering the freezing point, which is why salt is used on icy roads, so that the freezing point is lowered below the prevailing temperature and the salty ice melts.

Merilyn used the manufacture of ice-cream to encourage the children to explore the way variations in the process could affect the outcome. The idea of a 'control batch' was discussed and a batch was made using a basic recipe (see Figure 4.1). The ingredients were mixed in a small jar with a lid and this was placed in a tub of salted ice. The class discussed why these ingredients were used (drawing on their knowledge of the commercial process), and described the outcome in terms of colour, flavour and texture etc.

Merilyn's class discussed how the basic recipe could be altered, keeping in mind the processes they had seen at the factory. They wrote down on the worksheet how they would alter the process and what they thought the effect would be.

The variations that were made included:

- altering the proportions of the ingredients;
- heating the mixture prior to freezing (the factory heats the mixture to 80°C and then freezes it);
- whipping the mixture well, prior to freezing, to put air in it;
- freezing the mixture more slowly;
- freezing, thawing and re-freezing.

In another lesson the class focused on how the procedure could be optimised and further questions were raised. Some children were interested in finding out how the ice-cream would taste (and freeze) if they used different sorts of milk, like skim milk or sweetened condensed milk. They were keen to find out other ways in which milk could be changed.

(Adapted from Tytler & Costa 1998)

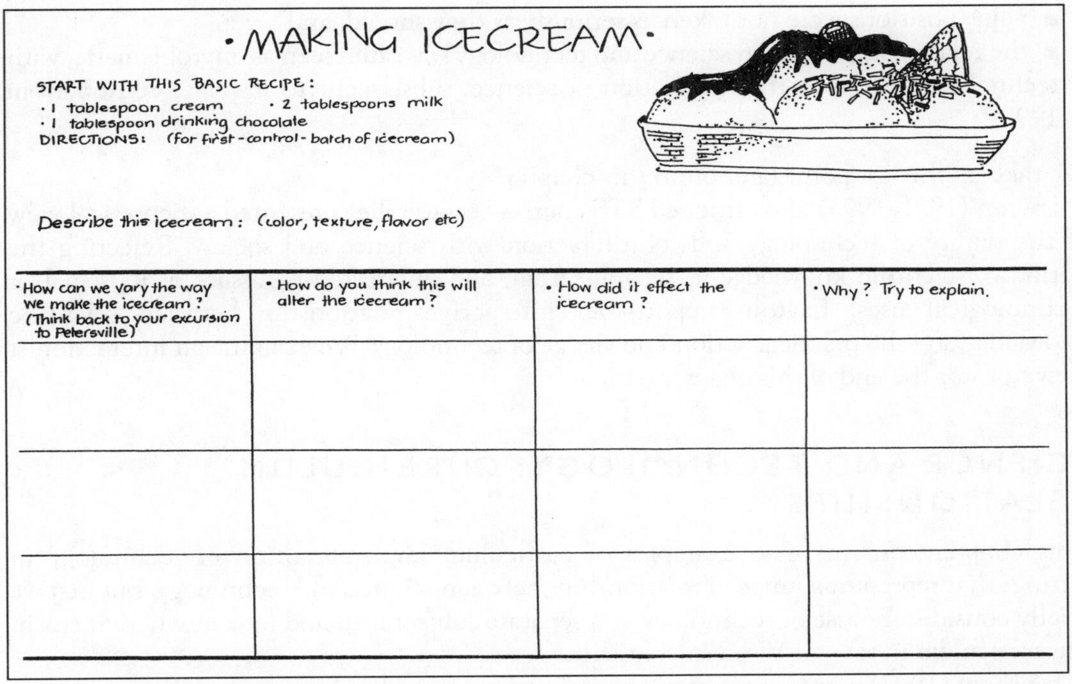

• MAKING ICECREAM •

START WITH THIS BASIC RECIPE:
- 1 tablespoon cream
- 2 tablespoons milk
- 1 tablespoon drinking chocolate

DIRECTIONS: (for first-control-batch of icecream)

Describe this icecream: (color, texture, flavor etc)

• How can we vary the way we make the icecream? (Think back to your excursion to Petersville.)	• How do you think this will alter the icecream?	• How did it effect the icecream?	• Why? Try to explain.

Figure 4.1 Designing an ice-cream production process

JOURNAL ENTRY 4.2 *Making ice-cream*

Merilyn ran this unit prior to technology being declared a distinct key learning area.

Is there a clear demarcation between the scientific knowledge and the technological knowledge in the ice-cream activity?

Would it be productive to separate the two areas in this activity?

Marilyn's excursion to the local ice-cream factory provided a meaningful context for science learning, and her unit is a good example of the STS (Science–Technology–Society) approach, which was at its peak in the 1970s and 1980s. One reason for the success of the STS approach was its emphasis on local aspects of science knowledge as it applied in context, and its recognition that social needs and political judgments determine which technologies are developed and used. With the STS approach, content and pedagogy emerged which made school science more relevant and meaningful, particularly for girls. However, there was concern about the way the STS courses were implemented because:

- technology was treated as an object of study, rather than as a set of knowledge and skills in its own right;

- value positions were not taken as seriously as they should; and
- the relationship between science and technology was often seen as unproblematic, with technology treated as the application of science, subservient to it (Hart & Robottom 1990).

We take up the last point later on in this chapter.

Layton (1991, 1993) also criticised STS courses because they presented a theoretical view of the nature of technology and its interaction with science and society. Rejecting the notion of scientific knowledge as separate from, and unaffected by, issues of its social or technological uses, Layton supports an interactive relationship between scientific knowledge and the practical action knowledge of technology. We examine an interactionist view towards the end of this chapter.

SCIENCE AND TECHNOLOGY: CURRICULUM RELATIONSHIPS

This chapter contains case examples of curriculum implementation of technology in settings that represent a range of relationships between science and technology. But first we briefly consider the rise of technology as a separate subject area and look at where it sits in the curriculum in several Western countries.

Fensham (1997) points out that in countries where there have been national projects, the socio-politics of curriculum decisions have discouraged the integration of science and technology. For example, in 1988, the government of England and Wales ordered a national curriculum for the compulsory years of schooling which resulted in a science curriculum consisting of the four strands—biology, chemistry, physics, and processes and skills. Technology was designated as a separate subject field: Design and Technology (D&T) and has its own Orders. Ritchie (1995) describes case studies in primary D&T which indicate that many primary teachers continue to plan and implement the curriculum in an integrated way, despite the national curriculum and its subject-based framework. In Australia and New Zealand national curriculum projects also specify technology as an independent subject area. In the United States technological literacy is the main aim of the 'Technology for All Americans' project which has recently developed a rationale and structure for the study of technology, and is producing standards for K–12 which will indicate what children need to know, or be able to do, in order to be technologically literate. Scriven (1987) also regards technological literacy as a main objective of technology education, and stresses the importance of citizens being able to use, adapt and evaluate new technologies.

In other countries integration rather than specialisation occurs. For example in Canada (where there has not been a national curriculum project) STS-influenced curricula exist in some provinces for the compulsory years of schooling. In the elementary years in Ontario, a 'broad program area' integrates the three fields of mathematics, science and technology in *The Common Curriculum* (Orpwood 1995). Similarly, in the Scottish 5–14 curriculum *National Guidelines for Environmental Studies* (SOED 1993) science and technology come together as part of a cluster of subjects—science, social subjects and technology, health and information technology.

In his writings about technology, which include an historical perspective, Gardner challenged the commonly held assumption that technology grows out of science (Gardner 1990; 1994a,b; 1995) by identifying four views of the science–technology relationship:

1. the materialist view (technology as a necessary precursor to science);
2. the idealist view (technology as applied science);
3. the demarcationist view (science and technology are independent fields); and
4. the interactionist view (scientists and technologists working together).

We have explored these views from the classroom perspective and extended the interactionist view to include a symbiotic approach, where science and technology not only interact, but are dependent on each other (Jane & Jobling 1995; Jobling & Jane 1996). The whole of Chapter 5 is devoted to an exemplary case study of this approach.

SCIENCE COMES FROM TECHNOLOGY (MATERIALIST VIEW)

Let us consider two curriculum examples of the 'materialist' view which involves toys.

1. Tinkering

Tinkering is one way technology can be used to stimulate science learning. Tinkering provides opportunities for children to discover the workings of mechanisms such as door locks, water taps etc. and household appliances such as toasters (for safety, cut the power leads off first). Tinkering, in order to find out how things work, was suggested by the McClintock Collective (1988) as a strategy which encourages the participation of girls. This may not be the case for very young girls as you will see in Chapter 11 where the gendered nature of tinkering is critically examined.

Through tinkering children can engage practically in the exploration of the design and construction of technological inventions in the 'real world'. Such experiences can contribute to technological literacy. Below is an opportunity for you to do some tinkering.

JOURNAL ENTRY 4.3 *Tinkering with toys—what's inside?*

1. Find a toy which moves that can be disassembled to see the moving parts.
2. On a separate sheet of paper write out the jumbled instructions (below) then cut them up.
3. Arrange the instructions in order before beginning to take the toy apart.
4. As you disassemble the toy try to identify:
   (a) the bits that make the toy work (these are the **functional** bits).
   (b) the parts that hold it together (these are the **structural** parts).
5. After trying to reassemble your toy write down in your journal your feelings about this tinkering experience.

continues...

Instructions (not in correct sequence)
- Take the toy apart using the correct tools remembering that *you* must put it together again. (Try to remember the *order* in which you took things apart).
- Choose a toy you would like to explore inside. My toy is ....
- Decide what tools you will need to use and collect them.
- After you have explored your toy long enough, put it back together again. How successful were you at doing this?
- My toy does the following things...............
- Can you guess some of the parts you might find inside? List the parts and what you think they do.
- After opening up the toy, draw what you see and label as many parts as you can.
- How many different materials were used in the toy?
- Think of some reasons why these different materials were used.

2. Designing, making and appraising a toy or gadget

Another way toys can be used in primary technology involves children designing and making their own toys. In a case study of a Year 5 class designing toys, a primary school teacher (Wendy) consulted with a final year engineering student (Peter) and together they set a technological task which required the children to make a toy that moved, or had moveable parts. It was expected that every child would produce an individual toy, although they could choose to work in pairs.

At the start of the unit Peter brought in a range of household tools, such as pliers, wine bottle cork remover etc. and models showing cogs and wheels which changed direction. The children were each given a tool to examine carefully, with the purpose of developing their understanding of how the various parts worked. After discussing their tool with a partner, the children sketched it, labelled the various parts indicating the direction of movement, and wrote an explanation of how the tool worked. Through discussion of how the various parts functioned separately, and how they related to the operation of the whole device, the children gained ideas for the design of their toys.

In subsequent lessons the children role-played being cogs and wheels, watched a video of a steam engine showing its operation, and observed a model steam engine which Peter set going. The steam engine demonstration was intended to get the children to think about steam as a source of energy, and for them to observe the cam in action. The gendered nature of these examples was taken into account due to Wendy's commitment to a gender-inclusive curriculum, as she explained.

'A lot of the examples we had initially were little steam trains and what might be considered to be masculine-type toys. Then we looked at the idea of dolls and the moving parts they would have in them. Also giving the children the option of making gadgets to do something—fun gadgets. Again you are trying very consciously to provide a learning experience to appeal to everybody.' (Wendy)

CHAPTER FOUR 63

Following on from these introductory activities the children were challenged with the following design brief (Figure 4.2) which contained the criteria for their toy or gadget construction.

1. The toy/gadget must have *at least two* different moving parts.
2. You should make a clearly labelled and drawn plan, preferably to scale.
3. You should list the materials you intend to use and how they will be joined.

Figure 4.2   Toy or gadget design brief

JOURNAL ENTRY 4.4   *Using the design brief to evaluate the product*

In the previous chapter you considered the general format of a design brief. Examine the design brief above. In your journal explain how this design brief could be used to evaluate the completed products.

Peter encouraged the children to think up lots of ideas for their toy through brainstorming. They were expected to draw several design sketches and then select one design by applying the OFFER guidelines (see Figure 4.3).

Objective:	Write down things you are trying to do.
Function:	The things it does
Factor:	Men & women—people to help you
	Money, also materials
	Machines—what will I need to make it?
	—will I need scissors or other tools?
	Methods—what sort of methods are you going to use to make it?
	Minutes—how long will I take to do make?
Effects:	
Restrictions:	Rules or requirements found in the design brief.

Figure 4.3   OFFER guidelines

Look at the following designs Tom drew before he started making his toy boat called a 'fishing smack'. (Figures 4.4 and 4.5).

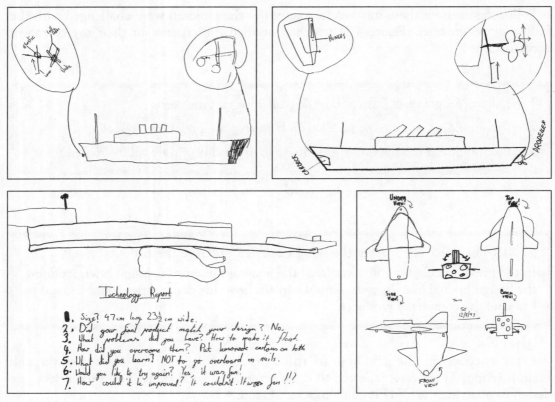

Figure 4.4   Tom's designs and report

Figure 4.5   Tom (left) with the materials needed to finish his boat, and (right) making the toy

We need to know more about how children generate their designs, and how teachers of technology can assist children to develop their design ideas. The ideas presented in Chapter 10 (which focuses on design) can help us here.

Wendy insisted that the children compare their final product with their original design, and then justify any modifications made. Tom made changes to his boat during the making phase as he tested it and overcame many problems during the process.

'The bottles move to act as rudders. Even that isn't my original design. I changed it as I went.' (Tom)

Although Tom was an opportunist in finding a hinge for his boat, he also showed perseverance to complete the product which was a bath toy for his baby cousin.

Mary also had a real purpose for making her toy doll because it was to be for her younger sister. She was very creative incorporating a cam to operate the moving parts of her doll. Mary understood how a cam works and set about making one to move the doll's legs. The introductory activities such as the role-play were important in developing her understanding of gears and cams. Below Mary explains how a cam works and she talks about her class experiences (see Figure 4.6).

'It's (the cam) like a pear shape and it can be made out of cardboard or metal and you use it to attach rods or anything really to it, and with mine when you turn the arms it operates the cam inside it which is attached to the legs. I could see how the cam worked when we were watching some videos and parts with gears. If you look at the different types of gears you know what happens when you turn the handle and we made chains in a group just like gears. With our class we had to put one hand on our hip and one hand next to us and they were like teeth.' (Mary)

Mary's understanding of the task, and her feelings about it, are revealed below as she describes the total process, including the design, and how she persevered with making the cam (see Figure 4.7).

'I had three different pictures at first and then I came up with this one which was going to be the main one. I had some problems and the materials, I wasn't sure what I was going to use. Then I came up with using cardboard and I glued the rods and it was OK and it didn't turn out to be exactly the same, but it's very close. I like having the chance to use the tools and making things and learning about the different parts of things.' (Mary)

Figure 4.6 Mary explains to a student teacher about the cam which makes her toy move

Figure 4.7 Mary is making her toy doll with moving parts

When designing her clown, Melanie initially thought of one design but then chose to make a new design (see Figure 4.8). She was keen to include very simple instructions for the method so that someone else could clearly follow them. The detailed labelling on her design drawing arose from her concern for clear directions. In Figure 4.9 she is in the making stage of a technological process.

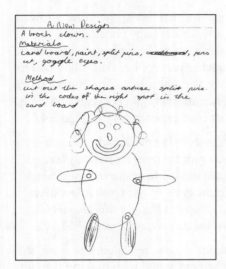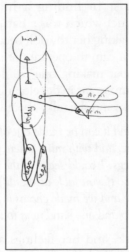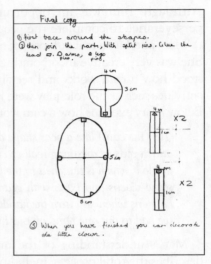

Figure 4.8   Melanie's design of her toy

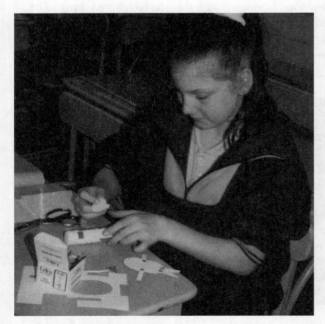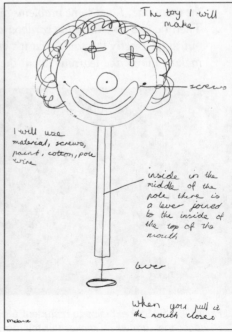

Figure 4.9   Melanie working on stage 1 of her project

In the unit on toys just described the technology (tools activity) preceded the science and was not subservient to it. This 'materialist' view can also be seen in technological inventions made by craftspeople prior to the scientific theories being developed. Examples include the light microscope, a very important technological development which led to scientific discoveries relating to cellular structure, and clock technology invented in the mid-14th century.

Materialist view of technology education involves:
1. Considering the technology before the science.
2. Focusing on the idea which guides the production of the product.

Assumptions
1. Technology is a way of revealing the world: 'It is a certain way of experiencing, relating to and organising the way humans relate to the natural world' (Ihde 1983, p. 29).
2. Technology is prior to science.
3. Technology is not necessarily subservient to science.
4. The idea is pre-eminent, guiding and directing the action (Carr & Kemmis 1986).

Teacher's role is one of:
- Introducing technological inventions or artefacts, such as tools, so that children gain first-hand experiences of the technology.
- Using these experiences to develop the children's understandings of the scientific principles that help explain how these technologies work.
- Assisting children to brainstorm their ideas and to apply the OFFER guidelines.

TECHNOLOGY AS APPLIED SCIENCE (TAS VIEW)

Technology as the application of science is a common perception portrayed in the media, and government policy documents, where science and technology are frequently coupled together. This linkage and order suggest that scientific knowledge is essential for technologists to create artefacts. Examples from history illustrating this belief are the electrical and nuclear power industries, which have science foundations.

Technology as applied science (TAS) is the view adopted in the New Zealand science curriculum statement (Ministry of Education 1995). Jones and Mather (1996) critiqued technology being taught in science education, and argued that in this context the science subcultural effects on both teachers and students must be recognised. Moreover, treating technology as embedded in science is problematic because other forms of knowledge essential for technology are overlooked.

Designing and building nesting boxes

Jobling and Jane (1996) describe a unit Wendy taught where the technology was an application of science. For the nesting box activity the success of the technological product depended on the children doing science prior to doing the technological activity. An environmental problem was the basis for the nesting box activity. The council had chopped down several old gum trees with hollows because of possible danger to children if the trees fell down. A class discussion of the tree-lopping program's effect on reducing the available nesting sites in the schoolground led to a unit which linked science with technology.

The children designed the nesting box after gaining information by observing the birds' behaviour and searching the literature on the topic. Designs were drawn and the properties of various materials (e.g. timber with water-resistant finish) considered prior to making the products. The designs were evaluated when problems arose with joining of the materials and the designs were modified accordingly. Careful thought was given to final placement of the nesting boxes to suit the birds' needs and the areas within the schoolground. Permission was gained from the council for parents to place the completed nesting boxes in the trees (see Figure 4.10).

Figure 4.10   The children's letter to the school council

Figure 4.11   Designing, constructing and placing nesting boxes is an important technological learning experience

Designing, making and appraising bird feeders

Teaching technology requires that children, as designers, meet real needs. The following year Wendy's class observed the birds in the schoolground near the nesting boxes. The children asked if they could make bird feeders to provide food for the possible inhabitants of the nesting boxes. Wendy agreed, so the groups set about designing bird feeders.

In response to Wendy's question: 'What do we need to think about before we make our bird feeders?' the children carried out simple research activities such as:

- making observations of birds in the schoolground to produce a class list of native and foreign birds;
- searching the literature (books and pamphlets) for feed recipes for native birds;
- examining existing designs (e.g. in books or outside a nearby hospital) which they might change; and
- choosing specific feeder designs for specific birds.

Following a class discussion of the materials, joining methods and the size of the feeders, the children were then in a position to start drawing their designs as part of the technological process. Kylie's bird feeder (see Figure 4.12) was for the honey eaters which she had observed in the schoolground.

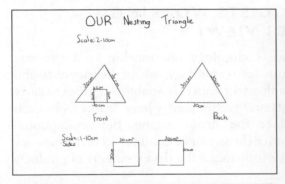

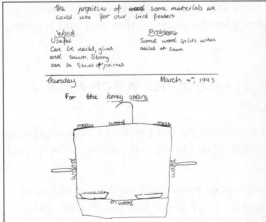

Figure 4.12   Kylie's design and final product for a bird feeder for the honey eaters

In both the previous examples (nesting box and birdfeeder) it was necessary for the children to develop their understanding of scientific concepts about specific birds' behaviour and food requirements. Based on this scientific knowledge, the children made appropriate decisions which influenced the designs for their products. The success of both technological tasks depended on the children's ability to transfer their science understandings to the technological context.

Technology as applied science view of technology education involves:
1. developing an understanding of the relevant science concepts.
2. applying these scientific concepts to solve a problem or to meet a perceived need by designing and making a product.

Assumption
Scientific knowledge is required to make the artefacts.

Teacher's role is one of:
- Planning a project whereby the children carry out 'scientific research' before they design and build their products.

SCIENTISTS AND TECHNOLOGISTS WORKING TOGETHER (INTERACTIONIST VIEW)

Proponents of this view regard the science–technology relationship as a two-way interaction. Science often provides a purpose for technology, whilst instruments and other inventions designed and made by technologists frequently enable scientists to carry out their investigations. In the past scientists and technologists have learnt from each other and have worked together to produce the steam engine, Bell's telephone, pneumatic pistons and energy-efficient machines (Fensham & Gardner 1994). These are examples of discoveries in science which have influenced the developments of products and vice versa.

A curriculum example of the interactionist view is a unit on designing and making a torch. Below Shirley (a primary teacher) describes why she planned this unit and her own experience of constructing a torch.

TECHNOLOGY TASK: DESIGN, MAKE AND APPRAISE A WORKING MODEL OF A TORCH

This case example illustrates the importance of constructing the product yourself prior to challenging the children with the task. Shirley began by tinkering and pulling a torch apart and drawing up her design (see Figure 4.13).

> 'Successfully building the torch to my own specifications proved to be quite a challenge. I bought a cheap $2.00 plastic torch and pulled it apart to see how it was constructed. All of the components were clearly visible and it looked quite easy to construct. However, using cardboard as the material for the container meant that the batteries were not held in tightly

enough to make a good connection with the globe. If I pushed too hard on the batteries the globe popped out and broke the circuit. In the end I had to glue the light globe holder into the torch.' (Shirley)

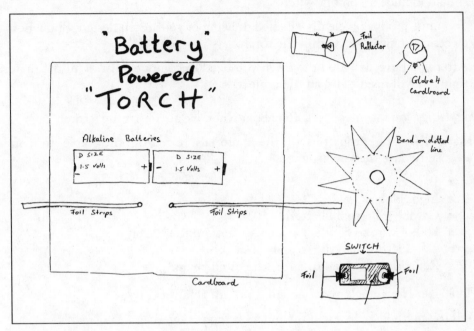

Figure 4.13   Shirley's design for her torch

Shirley agrees with the ideas of Black and Harrison (1985) (cited in Fensham & Gardner 1994, p. 166) when they point out that science and technology should not vie for a master and servant relationship. Rather they should be seen in a more balanced relationship, with each offering solutions to societal problems which will hopefully generate the purpose and motivation to learn.

Shirley decided to use a technology unit to increase children's understanding about science principles relating to energy because energy is a difficult concept to teach. She believes that a technology unit will involve the children in 'hands on' exploration of materials as they go about the task of making their torch work. Technology provides the catalyst for children to 'need' to understand about electricity and circuits. The science comes about quite naturally as the children go about their design and construct stages. Then, if any faults occur, and the globe does not glow in the final product, the children will need to solve the problem by looking to make sure that they have constructed the torch along sound scientific guidelines.

Torch checklist

Using Shirley's design brief below (Figure 4.14) the children can design, make and appraise a working model of a torch.

> 1. You and your partner are to design and build a working model of a torch. The torch must be able to cast a beam of light three metres onto a wall. Your design should include an on/off switch.
> 2. You must produce a clearly labelled drawing of your torch design. Show how the electric circuit works within the torch.
> 3. After you have designed the torch you need to compile a list of all the materials that you will need to finish the project.
>
> Some things you may need to think about when we are designing a torch.
>
> 1. **Materials:** The properties of the materials, e.g. Are they able to conduct electricity? Are they strong enough?
>   - Batteries
>   - Light globe
>   - Cardboard to hold batteries
>   - Foil or wire to conduct electricity around the circuit
>   - Conducting material to make a switch
>   - Reflecting material to direct the torch beam.
> 2. **The size:** How much material will we need? (No wasting).
> 3. **Joining methods:** How will the model be constructed?
>
> (Adapted from material prepared by Shirley Noble)

Figure 4.14   The design brief

Curriculum area: Science, technology, language.

Statement of problem: To design a working model of a torch.

Teaching style: Children working in pairs, teacher roving and asking questions to help focus children on 'how a torch works'. Are the batteries in good contact with the conducting material and the light globe? Have you checked that the light globe works or that the batteries are not flat?

Prior knowledge: Children should have had some experience with circuits, working with a range of materials and drawn designs.

Interactionist view of technology education involves:
1. Technology and science engage in two-way interaction.
2. Science often provides a purpose for the technology.
3. Technological developments can assist scientific discoveries.

Assumption
'Today there is no hierarchical relationship between science and technology. Science is not the mother of technology. Science and technology today have parallel or side-by-side relationships; they stimulate and utilise each other' (Franklin 1992, p. 33).

Teacher's role is one of:
- Planning a project whereby the science provides a purpose for the technology.

SYMBIOTIC SCIENCE–TECHNOLOGY RELATIONSHIP

We use the term 'symbiotic' to describe the position where science and technology interact in a mutually beneficial way (Jane & Jobling 1995). Consider bread making which is a popular context in primary schools, yet the role of yeast in the process is often downplayed. In New Zealand, where biotechnology is one of the seven areas in the national technology curriculum, France and Chambers (1997) found that when yeast is identified as the living component of the biotechnological process, and its physiology taught, the biotechnology program is enhanced.

'We can make bread'—development of a bun for a niche market

In the program developed by Megan Chambers (Ministry of Education 1997) an applied science view of technology was not adopted. During the science program for the Year 2 and 3 children they followed the scientific method and manipulated variables to answer the question 'What is yeast?'. The class developed a shared diary of its understanding of the scientific method and the physiology of yeast.

> *We know what a control is. It's one experiment that is normal so that we compare results. We have also learned that yeast is made up of tiny living bugs but what makes them rise or not rise is in the yeast mixture?* (France & Chambers 1997, p. 3)

The children's technological skills were developed as they 'identified and selected equipment, measured ingredients, activated the yeast, mixed and proved the dough, kneaded the dough and formed it into buns' (p. 3). The children were challenged to alter their basic bread recipe to suit their own specialty bun for their niche market (e.g. adults, babies, sick people etc.). A taste test evaluation of their buns occurred with each target group, and necessary alterations were made.

The science program contributed to the biotechnology unit 'We can make bread' by involving the children in the following science procedures and processes:

- Fair testing
- Use of a control as a comparison
- Quality-control procedures
- Scientific understandings (physiology of yeast)
- Scientific understanding (yeast is a living organism)
- Quantitative measurement
- Identification of the variable.

From records in the shared diary the following table emerged (see Table 4.1).

Table 4.1   Identification of science knowledge and technical skills that contributed to the development of technological knowledge

Science knowledge	Technical skills	Technological knowledge
Yeast is a living organism	Observation How to activate yeast	Identification of the optimum conditions for dough proving
Yeast requires sugar and warm water to activate and takes time	Measurement of ingredients Sequencing Following a recipe	Adaptation of a recipe for a niche market Allowing appropriate time for proving
Bubbles of gas are produced when yeast works	Observation Measurement and comparison	The proving period is limited Identification of ingredients and optimum conditions for proving
Yeast activity changes when you alter the substances	Identification of components Measurement	Varying the recipe according to specifications Quality control of the bread making process
Yeast works better with warm water than boiling or cold water	Temperature measurement Comparison of bubble production	Identification of the optimum temperature to activate the yeast
Scientific method only allows one variable to be changed at a time	Selection of the variable Keeping all other variables constant	Realistic predictions about the effect of changing the conditions or additives
A control is used as a comparison	Measurement of bubble height Description and conversion into a visual image on a histogram	Quality control Comparison of product against standard recipe

(France & Chambers 1997, p. 4).

> **JOURNAL ENTRY 4.5**   *How do we use yeast?*
> Why was it important for the children to know the scientific understandings relating to yeast?
> What was the advantage of keeping a shared diary or class book?
> How important was the idea of a niche market to the success of the unit?
> What advantage was there in not making the science component the major focus?

In this case, biotechnology was used as a stimulus for technology education. This unit is an example of a symbiotic relationship between science and technology, where the science of yeast made a strong contribution to the children developing technological knowledge and skills. This kind of science–technology relationship is further explored in chapter 5.

Symbiotic view of technology education involves:
Science and technology interacting in a mutually beneficial way.

Assumption
The science–technology relationship is a two-way interaction that can proceed in either direction.

Teacher's role is one of:
- Organising a project where the technology depends on the children's understanding gained from their science work, whose purpose is to feed into the technology task; or
- Challenging the children to design and make a technological product which is then used to facilitate their scientific investigations.

SCIENCE AND TECHNOLOGY ARE INDEPENDENT FIELDS (DEMARCATIONIST VIEW)

Not all writers view science and technology as being related. Scriven (1985) and Fleming (1989) contend that technology is independent from science because technology has its own knowledge and skills.

> *Science is defined as the process and publicly accessible product of our attempts to describe, explain and predict natural phenomena. Technology is the systematic process, and the product, of designing, developing and maintaining and producing artefacts* (Scriven 1985, cited in Rennie 1987, p. 122).

Consistent with this view of technological knowledge as distinct from scientific knowledge is Gunstone's (1994) argument that engineering can be considered a different way of knowing from physics.

One of the early attempts to design a technology curriculum in this country began in 1987 in Western Australia when curriculum modules were developed for teaching technology to primary school children (Kinnear, Treagust & Rennie 1991). In this project technology was recognised as problem solving to meet human needs, and defining it in this way sets technology apart from other areas, such as science, which also involve problem solving. The 'human needs' context provides opportunities for children to question and evaluate the technological development in terms of its possible social costs and consequences, including environmental impact.

A problem-solving activity which relates to the schoolground environment was carried out by Year 1 children. The science of materials was not specifically introduced to the children.

SUMMARY OF PROBLEM-SOLVING ACTIVITY USING DESIGN, MAKE AND APPRAISE STRATEGY

Year 1 with Year 6 input

A. Introduction
The Year 1 children observed the playground which was littered with rubbish. Big black ravens had pulled the rubbish from the bins after the lunch break when all children had returned to their classrooms.

continues…

B. *Design brief*
- The problem was discussed back in the classroom and written on the board. 'The birds take rubbish from our bins and litter the playground with it.'
- The task was decided upon.

To design an original bin top which will allow the children to place their rubbish in, but which will prevent the birds from taking it out.

- Constructional constraints:
  Safety to be paramount (no sharp edges on bin designs, for example).
  Miniature bins could be used.
  Bins and lids must be light enough to be handled by children.
  Time set for construction: one week.

- Methods of appraisal were discussed.
  There would be presentations to the class, followed by class input re design and possible alterations suggested.
  The teacher would observe skills and attitude development by observing 'how well they worked together, and how thoughtfully they designed their models'.

C. *Teacher made her own model in out-of-school time*

D. *Class initial planning with children discussing in groups*
Pointers given to help stimulate discussion:
– Again observe rubbish in the playground.
– Observe hopper into which the bin needed to be emptied. (Consider height of hopper and need for light bins.)
– Discuss present types of bins available. Would they be useful for outside? Why or why not?

E. *Children came together to share ideas. Problem and task redefined.*

F. *Planning and designing stage*
Worksheet type booklets distributed and discussed.
  1: Child's name. Names of others working with the child.
  2: Statement of problem. Space for illustration. Statement of task.
  3: Design drawn (can be redrawn and altered up to four times).
  4: Things I will need (can be written or drawn).
  5: What my finished invention will look like.
     —Children then hand their completed booklets to the teacher for checking.
     —Children gather required resources.

G. *Children construct their bin lids* (in some cases entire bins)

H. *Appraisal and presentations of bins to class* (and later to Year 6 also, in order to get extended input).
  Children then had open discussion directed by teacher as to usefulness, safety etc. of bins.

continues...

I. Unplanned extension
Year 6 children became interested and designed and made their own bins to solve the playground problem, and presented them to the Year 1 students and these were also discussed.

J. Assessment
Ongoing discussions with individual children took place. Observations of children's work, attitudes and skill development were made throughout the unit.

(Adapted from unit prepared by V. Smiley)

Demarcationist view of technology education involves:
Technology can be viewed as an independent system of thought and practice.

Assumptions
1. Science and technology have differing goals, methods and outcomes, and are independent (Gardner 1994a).
2. Technology is a unique way of thinking and is 'an autonomous realm of knowledge' (Lewis & Gagel 1992, p. 127).

Teachers' role is one of:
- Implementing a technological task by focusing on how to do or make things involving a systematic process of designing, developing and producing products.

SUMMARY

Teachers over the years have used technology as applied science to enrich their science programs and stimulate science learning. In this chapter the vignettes of classroom practice have illustrated that science and technology can interact in different ways to successfully enhance children's learning of scientific knowledge and technological knowledge. Therefore it can be productive for teachers to devise design briefs which challenge children to engage in activities which cover a range of ways that science and technology can relate to each other.

It has been shown that topics involving products such as toys, nesting boxes, bird feeders, and bird-proof rubbish bin lids are relevant contexts for technology learning. In the next chapter the context of the schoolground is used to explore a symbiotic approach to curriculum implementation of technology in more depth. This view is typical of the trend identified by McGinn (1991).

> *The relationship between technology and science has changed markedly over the millenia, from virtual isolation and mutual independence to generally close association and mutually beneficial interdependence. What is indisputable, however, is that during the past 130 years, that relationship has been one of progressive symbiosis* (McGinn 1991, pp. 26–7).

REFERENCES

Black, P. & Harrison, G. (1985) *In place of confusion: technology and science in the school curriculum*, National Centre for School Technology, London.

Carr, W. & Kemmis, S. (1986) *Becoming Critical: Education, Knowledge and Action Research*, Falmer Press, London.

Fensham, P.J. (1997) 'Continuity and discontinuity in curriculum policy and practice: case studies of science in four countries', in the Proceedings of the 8th IOSTE Symposium, 1996, Edmonton, August, pp. 32–6.

Fensham, P.J. & Gardner, P.L. (1994) 'Technology education and science education: a new relationship?', in *Innovations in Science and Technology Education*, D. Layton (ed.), UNESCO, Paris, (pp. 159–70).

Fleming, R. (1989) 'Literacy for a technological age', *Science Education*, 73 (4), 391–404.

France, B. & Chambers, M. (1997) 'The science of yeast: more than a silent partner in biotechnology'. Paper presented at the inaugural Technology Education New Zealand Conference, Christchurch, October.

Franklin, U. (1992) *The Real World of Technology*, House of Anansi Press, Concord, Ontario.

Gardner, P. (1990) 'The technology–science relationship: some curriculum implications', *Research in Science Education*, vol. 20, pp. 124–33.

Gardner, P. (1994a) 'Representations of the relationship between science and technology in the curriculum', *Studies in Science Education*, vol. 24, pp. 1–28.

Gardner, P. (1994b) 'The relationship between technology and science: some historical and philosophical reflections. Part 1', *International Journal of Technology and Design Education*, vol. 4, pp. 123–53.

Gardner, P. (1995) 'The relationship between technology and science: some historical and philosophical reflections. Part II', *International Journal of Technology and Design Education*, vol. 5, pp. 1–33.

Gunstone, R.F. (1994) 'Technology education and science education: engineering as a case study of relationships', *Research in Science Education*, 24, pp. 129–36.

Hart, E. & Robottom, I. (1990) 'The science–technology-society movement in science education: a critique of the reform process', *Journal of Research in Science Teaching*, 27 (6), 575–88.

Ihde, D. (1983) *Existential Technics*, New York Press, Albany.

Jane, B.L. & Jobling, W.M. (1995) 'Children linking science and technology in the primary classroom', *Research in Science Education*,. 25 (2), pp. 101–201.

Jobling, W.M. & Jane, B.L. (1996) 'Exploring science-technology relationships from the classroom perspective'. *Australian Science Teachers Journal*, 42 (2), 37–9.

Jones, A. & Mather, V. (1996) 'Technology in science education: implications for teaching and learning'. Paper presented at Australasian Science Education Research Association, July 1996, Canberra.

Kinnear, A., Treagust, D. & Rennie, L. (1991) 'Gender-inclusive technology materials for the primary school: a case study in curriculum development', *Research In Science Education*, 21 (21), pp. 224–33.

Layton, D. (1991) 'Science education and praxis: the relationship of school science to practical action', *Studies in Science Education*, vol. 19, pp. 43–79.

Layton, D. (1993) *Technology's Challenge to Science Education*, Open University Press, Buckingham.

Lewis, T. & Gagel, C. (1992) 'Technological literacy: a critical analysis', *Journal of Curriculum Studies*, 24 (2), pp. 117–38.

McClintock Collective (1988) *Getting into Gear*, Curriculum Development Centre, Canberra.

McGinn, R.E. (1991) *Science, Technology and Society*, Prentice Hall, Englewood Cliffs, New Jersey.

Ministry of Education (1995) *Technology in the New Zealand Curriculum*, Learning Media, Wellington.

Ministry of Education (1997), *Towards Teaching Technology: Knowhow 2 Book 2*, Ministry of Education, New Zealand.

Orpwood, G. (1995) Assessment of Science and Technology Achievement Project. Working Paper No. 1. *Science and Technology Standards for Ontario: A Model for Development*, Science Education Group, Faculty of Education, York University, North York, Ontario.

Rennie, L.J. (1987) 'Teachers' and pupils' perceptions of technology and the implications for curriculum', *Research in Science and Technology Education*, 5 (2), 121–33.

Ritchie, R. (1995) *Primary Design and Technology. A Process of Learning.* David Fulton Publishers, London.

Scriven, M. (1987) 'The rights of technology in education: the need for consciousness-raising'. Paper for the Education and Technology Task Force, Ministry of Education and Technology, Adelaide.

SOED (1993) *National Guidelines for Environmental Studies 5–14*. The Scottish Office Education Department, London.

Tytler, R. & Costa, M. (1998) 'Making ice-cream: incorporating an industry visit into the primary school curriculum', in *Enhancing Technology and Science Education through School-community Links*, B. Jane (ed.), Faculty of Education, Deakin University, Victoria.

ACKNOWLEDGMENTS

Thank you to V. Smiley, W. Jobling and S. Noble who provided the examples in the cases included in this chapter.

Chapter Five

A symbiotic approach

INTRODUCTION

Teacher: 'During the term, what did you enjoy most about your project work?'
Kelly: 'The part I liked best was making the bug house and actually using it.'
Barry: 'I liked catching the bugs and making all the things. I liked learning about slaters and looking up books to find out information.'

JOURNAL ENTRY 5.1   *Is this science or technology?*
What do you think these children have been doing in this project?
Did this project involve science or technology?

The case study in this chapter highlights how a primary school teacher thought about, and planned for, the project work these children were engaged in.

In Chapter 4 you looked at the relationship between science and technology. Many curriculum developers and teachers have also given thought to this relationship. For example, in South Australia, Keirl (1998) argues strongly for the separation of technology from science, claiming 'that our field has nothing of quality to gain from any more than a passing acquaintance ... with science' (p. 5). In contrast, Stead and Witt (1998) promote ways of enhancing the teaching and learning of technology through the application of an integrated curriculum. They draw on their own experiences of using integrated approaches with primary students, and that of their colleagues, such as Kathy Paige's (1998) unit of work on gears, which integrates the three areas of technology, science and mathematics. Paige retains the integrity of each area of study by highlighting the scientific, technological and mathematical ideas related to the activities centred on gears.

In the primary school setting, a symbiotic approach, where science and technology interact in a mutually supportive way, is very appropriate (Jane & Jobling 1995). This chapter features a case study of Year 5 children working with their teacher on a unit investigating small animals using this approach. The children produced technological products which were designed to facilitate their study of small animals which were found in the school environment. This unit was designed and implemented by the teacher, Wendy Jobling, and it is a good example of how the two key learning areas—science and technology—can be linked in a symbiotic way. Wendy described her intentions regarding this unit as follows.

> 'The unit we have just been working on was one where I wanted to incorporate science and technology studies, without the technology studies actually coming from the science. So in this case the technology work was assisting the science work the children were doing, and we were making use of the school environment. The children were required to look at trees and bushes within the school ground and then look at the dependence of the animals, or the interdependence probably, of the plant and the animal. What the children then had to do was to design something to catch an animal for closer study and then somewhere to actually house the animal that they had caught. We're talking about small animals such as slaters and millipedes. They had to consider how they were going to treat these animals. It had to be in an ethical way and they were learning about them, perhaps conducting simple experiments, but those experiments were not going to harm the animal in any way.' *(Wendy)*

You can see that Wendy appreciates that technology can be a stand-alone curriculum area in its own right. At the same time she recognises that there are advantages for students when they participate in a curriculum unit which links technology with the science curriculum area.

Research in New Zealand by Jones and Carr (1993) has shown that classroom practice which embraces technology is very much influenced by how teachers perceive technology. In the next section the focus is on Wendy's perceptions of technology, which are societally embedded.

THE TEACHER'S PERCEPTIONS OF TECHNOLOGY

Wendy's perceptions have been influenced by several factors, including her readings as part of her professional development activities. Below she talks about her view of technology and its status in our culture.

> 'One of the problems here is that our culture is one which sees technology as being a second class intellectual pursuit or activity, and I think it can be traced back to things like technical schools before the restructuring. If you went to a technical school you went because you weren't very academic … In Peter Ellyard's article, he's looking at the situation in Germany where people who go to the equivalent of TAFE (technical and further education) colleges, it's perceived to be different but not inferior. You see the Germans have got this culture, if you look at an advertisement for Mercedes Benz cars, the name of the designer is in the

advertisements. When do you ever see "Oh, this is the designer of the latest Holden Calais". They might say it is an Australian-designed car but they won't actually give public acknowledgment to the technical side, who developed what; so I think that is a problem, because if we are going to value it and change our ideas, we need to look at having a different perception of what technology is. One problem is that technology again is looked at as part of something that's not quite elite or valuable, even though we can't function without technology around us even in the simplest forms.' *(Wendy)*

In the transcript below you can see that Wendy views technology as sitting well in an integrated curriculum. She appreciates that there can be a range of solutions to a particular technological task, with children generating several designs and then selecting suitable materials with which to make their products.

'I think technology studies is a very worthwhile program and it can be very well integrated. I was talking to the children about making a pencil case. "In the past I would have got you all to cut something the same size and they all had to have a sliding lid and you have 20 identical products." And someone said "Oh, my brother made that pencil case you have just described". And I thought horror of horrors, obviously in some schools they are still making them. I was describing a pencil case I think my brothers made 20 years ago. Now if you are looking at the idea of a technology studies framework curriculum you would say to the children, "You have a lot of pencils, protractors, compasses and pens to carry around. What your design brief is, is to make a suitable container for them. You might want to make them out of wood". Then you might do different activities with the children in that situation, learning to saw pieces of wood, learning to glue, learning to use different tools and pieces of equipment, but they would do the design. Or some children might say "I want to use a sewing machine and make one out of fabric", and they would look at the different fabrics that were available.' *(Wendy)*

Wendy has a sound understanding of the range of ways science and technology relate to each other. In the next transcript she refers to her reading of the Victorian Technology Studies Framework (Maruff & Clarkson 1988) curriculum document and Paul Gardner's (1990, 1993) extensive writings on the science–technology relationships. Wendy has a knowledge of various technological inventions in the past which have preceded the science ideas.

'I had read the technology studies framework as a start. They were given out and we were asked to read them, so I had read them and I had my own copy of the science framework. I didn't have my own personal copy of the technology studies framework, but the science framework also has a reference to it, which is another interesting aspect, and this is what Paul gets into. Does technology come from science or vice versa, or are they both separate things? I think you can have instances of all sorts of things and his article is well worth reading on this because he looks at it from different positions. The science framework mentions technology and they have a viewpoint that technology comes from science and there are so many instances that you could quote where the science has come from the technology. Even in a lot of work in astronomy you can find the link back to the development of clocks, which was a technological development—the clockmakers—and there are lots of other instances you could even give in medical research, where because of the technology of microscopes and so on, they have been able to make discoveries, science-type discoveries. Blood circulation is another ... the child got the idea that the human body would have this pump concept, from several pumps, so again the science came from the technology.' *(Wendy)*

PLANNING THE CURRICULUM UNIT

In her curriculum planning Wendy deliberately chose to link technology with science in this unit. Setting the technological task in the context of the science work the children had experienced previously was one reason for the success of the unit. Relating the new task to prior activities is consistent with a constructivist orientation to learning. She organised the children into small groups of two or three, which the class decided would be formed by drawing names out of a hat. The unit was scheduled for one and a half hours, every Thursday afternoon, for eight weeks.

> JOURNAL ENTRY 5.2   *How much technology time?*
> Wendy prefers to teach science and technology linked from a children's science perspective. She has the whole class involved in the project at the one time, and believes it is important to have a set time per week for technology, otherwise it may be omitted in the already packed school program.
> What are your views?
> How much time do you think should be devoted to teaching technology?

Below Wendy explains how she planned the unit.

> 'When planning a unit such as this I looked at several aspects. What the children had done before, both this year, and trying to work out what they had possibly done in past years. So trying to come into the unit from where children have some background of knowledge that they can link their new knowledge to. I planned the unit from this stage, of what we were going to begin with. You are really planning a framework of incorporating your design, build and appraise, and your science areas. And you're allowing flexibility within that because, if you're going to try and stick to a rigid plan you're not going to allow for the directions that the children might be naturally interested in. My role in this unit is to set up that broad framework of what I want the children to achieve, the skills I want them to develop, the sorts of knowledge that I would like them to gain from the unit, and then to facilitate that. So being a facilitator is probably the key word.' *(Wendy)*

INTRODUCTORY ACTIVITY: TAKING ACCOUNT OF PRIOR KNOWLEDGE

At the start of the unit an introductory activity required the children to revisit the trees and shrubs they had used in science earlier in the year. This time the children had to carefully observe the leaves, and look for evidence of any animals that were living on them. This introductory activity was designed to link the technological task to the children's prior experience of observing selected trees and shrubs. Figure 5.1, a transcript of Wendy's talk to the class, indicates her expectations.

Plant study
1. Select a tree or shrub to study.
2. Are there different shaped leaves on the plant (adult/juvenile)?
   Closely examine a leaf from your plant. Consider its shape, size, texture. Can you suggest reasons for this shape? (e.g. How could the leaf shape help the plant?) Think about the advantages.
3. Bark—consider the colour, texture and its function.
4. Research the name of the plant, its origins, ideal growing conditions (design a possible experiment).

Animals in, on and under your plant
When you start looking for your animal around the tree that you have been studying, things about the animal to look at with the magnifying glass are:
- its legs (number of legs: spider—8, insect—6, slater—?)
- how it moves;
- what it looks like.

Figure 5.1   Introductory activity

THE TECHNOLOGICAL TASK: DESIGN, MAKE AND APPRAISE A SMALL ANIMAL-CATCHING DEVICE AND CONTAINER

Following on from their plant study, the children were challenged with the following open-ended design brief (Figure 5.2), which sets out the purpose of the technological task.

Design and make:

1. A device for safely catching a small animal such as a slater.
2. A suitable container to house the animal until you finish your study.

(It will then be returned to its original habitat.)

Figure 5.2   Design brief

> JOURNAL ENTRY 5.3 *Open-ended design brief*
> Why do you think Wendy selected a design brief which was so open-ended?

Prior to the children drawing their designs for the catcher and container, Wendy brainstormed with the class, asking the children what they needed to think about before they drew their designs. They suggested that the animals' welfare and safety should be a priority, and the design itself should be a clear drawing which showed the scale and measurements. The children were aware that their designs would be restricted by the available materials and their cost. In the next transcript Wendy describes the situation relating to the materials for technology at her school.

> 'Resources are a very important part of doing any work like this. We are making use of the natural resources within the school, and like many schools we're in the position of not having access to large amounts of money. So you are really focusing on using, as much as possible, recycled materials. You can have recycled cardboard scraps from parents who are in printing occupations, recycled PET bottles, pantihose. You do need to buy things such as masking tape, but in terms of actual expenditure, very little, as the only expenditure really for this unit is buying the masking tape.' *(Wendy)*

Figure 5.3 *A group of children making their animal enclosure using recycled paper and sticky tape*

Very early on in the unit Wendy emphasised the importance of groups appraising the effectiveness of their products. Accordingly, opportunities were provided throughout the unit for each group to communicate their ideas and 'work in progress' to the class, as Wendy indicates in Figure 5.4.

> When reporting back to the class you need to report on:
> - how well your catcher worked;
> - how well your bug house worked;
> - the animals you found and where you found them, on the tree or in the leaf litter around the tree.

Figure 5.4   Report to class

Most children (see two examples below) were capable of appraising how well their catchers and houses worked, and suggested improvements to their designs.

> 'With our catcher it worked well but the bugs could climb up the top and get out, and the slaters could also roll off. So next time I think we had better put something over the top and a flap to close the front.' *(Liz)*

> 'I thought ours were good. The bug catcher it didn't work but I think it would have worked better if we had plastic to make the house out of. If it was airtight, it would have worked better.' *(Kelly)*.

THE SCIENCE COMPONENT

Vegetation found in the animals' habitat was placed in the enclosures to provide food and shelter for the animals while they were housed temporarily. Once the animals were in the containers the children carefully observed the animals' structure and behaviour. The children had ownership of their work and were encouraged to pose their own questions, and were given flexibility in the way they went about seeking answers to these questions. Some children devised and carried out simple experiments, as Wendy describes in the next transcript.

> 'For instance, if they ask you "what does a slater eat?" and this was one question that children did ask me, I said "how do you think we can find out what it might like to eat?" You will find that by going down that sort of track children will say in this case, "we could give them a selection of foods we think they might like to eat and see which ones they go to eat" which is in fact what they did. Those teacher questioning skills are very, very important. You are a facilitator, not an answerer of questions, and not always in the role of a person who sets the questions that the children are going to answer. The children pose questions that I would never have thought of posing myself, and found the answers to them, and there is great

excitement. They call you over and say "look what we have found", and they are very enthusiastic about it because, again, they're the ones that own the work that they're doing.' *(Wendy)*

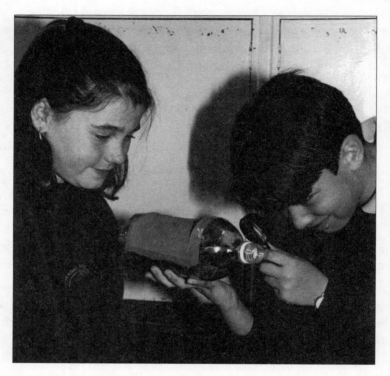

Figure 5.5   Children observing the animals in their temporary enclosure

Wendy recalls the classroom when the children observed the new baby slaters.

'The children have had many opportunities to discover things for themselves. For instance with the slaters, the thrills of delight when the slaters reproduced and all these little baby slaters underneath the mother or the parent slater, and they feel that they have made this incredible scientific breakthrough of finding that they actually have babies, even realising things such as that they are made up of segments, and observing how they start to walk and which legs move first.' *(Wendy)*

When talking about the methodology of integrating technology, Stead (1998, p. 1) identified that an integrated curriculum provides flexibility: 'Frequently the student will have been involved in deciding the topic, will decide parts of the topic which they will investigate, and will decide how they will illustrate what they have learned.' Consistent with this view, Wendy encouraged the children to present their findings and information in novel ways. Most reports revealed that the children had discovered science ideas relating to the structure and behaviour of the small animals they had found. Refer to the following report written by Liz and Emma (Figure 5.6).

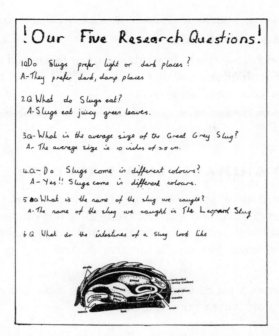

Figure 5.6   Questions Liz and Emma asked and found answers to

> **JOURNAL ENTRY 5.4**   *Structure and function*
>
> Question 6 'What do the intestines of a slug look like?', shown in Figure 5.6 above, arose from the girls' fascination with the slug's intestines. Suggest a follow-up activity which might lead them to a better understanding of the structure and function of an animal's intestines.

In his written report (Figure 5.7) Morris recorded the answers to his questions about the slater (or wood louse) that he observed.

Figure 5.7   The report by Morris

> JOURNAL ENTRY 5.5 *Plants and animals*
>
> Examine Lisa's report below which shows that she was interested in factual knowledge related to the slater's colour, size and diet.
>
> What else was she concerned about?

PLANTS AND ANIMALS (LISA'S REPORT)

Our grade has been studying some plants and the animals around the tree. Our group found a family of slaters and one tiny spider. We made a bug catcher and a bug house. In our bug house we had a lot of its surroundings where we had caught them. We had bark, wood chips, grass and leaves. Not one of our animals died!

Facts
Did you know … slaters are a dark grey! They generally move around in groups. A slater curls up in a ball to protect itself from its enemies.

Q/A on the slater
Q1. How does a slater move?
A1. The slater walks from the back legs onward.

Q2. What food does a slater eat?
A2. The slater eats wood (bark) and lichen (moss).

Q3. How many legs does a slater have?
A3. The slater has 14 legs.

Q4. How long are slaters?
A4. Slaters range from 5mm to 20mm.

Q5. Do slaters like the shade or the sun best?
A5. The slater seems to like the shade best.

Q6. How do you tell the slater's sex?
A6. The female has a yellow stomach and the male has two white dots and the rest extra white.

The comments below show that Katy learnt how slaters behave when confronted with possible predators, and that Neil enjoyed watching slaters move.

> 'When we were digging around to find some insects for our tree we couldn't really find any at first because we didn't know that when slaters try to protect themselves they curled up into little balls. We didn't know what they were. We thought they were little things that fall off trees, little seeds; and in the end we found some slaters and a little worm.' (*Katy*)

'I found out how slaters walk. They start moving their back legs first and when it gets to the second front leg the back legs start to move again.' *(Neil)*.

Refer back to the start of this chapter on page 81. Kelly's comment is indicative of the value of linking science and technology in the curriculum. Kelly liked using the technological product she made, and she could see a real purpose for participating in the technological task. The science component made the task an authentic one.

THE LEARNING ENVIRONMENT

Let us look more closely at the learning environment which Wendy created. The teaching strategies used reflect a social constructivist framework which recognises the importance of social, as well as personal aspects of learning. Wendy understanding that children construct their own meanings from the classroom context planned activities which allowed individual children to interact with new information in their own way.

In a study by McRobbie and Tobin (1997), the learning environment was interpreted from a social constructivist perspective using Tobin's (1993) Classroom Environment Survey (CES). Three dimensions of the CES: involvement, autonomy and relevance can be used as a basis for reflecting on the case study described in this chapter.

JOURNAL ENTRY   5.6 *Involvement*

'*A social constructivist perspective on learning highlights the role of active involvement in tasks associated with making connections between experience and extant knowledge ... The development of understanding by writing and discussion of ideas with peers is an essential element in learning and students should be given opportunities to speak and write about their science*' (McRobbie & Tobin 1997, pp. 197, 199).

In what ways did Wendy value student discourse as a source of learning?

JOURNAL ENTRY 5.7   *Autonomy*

'*A social constructivist perspective on learning suggests that students should have control over their own learning and construct meanings for their experiences in terms of what they already know at the time of learning*' (McRobbie & Tobin 1997, p. 199).

What opportunities did Wendy provide for the children to exercise autonomy? Were the children given a choice in terms of:

- how they went about problem solving?
- what they did at a given time?
- with whom to work?
- what to study?

> JOURNAL ENTRY 5.8 Relevance
>
> *The relevance of the curriculum is critical if children are to adopt an approach to learning that is centred on building understanding. 'Students' goals are influenced by the nature of academic tasks such that when they are more challenging, meaningful, and authentic or more interesting, important and useful students are more likely to learn with understanding'* (McRobbie & Tobin 1997, p. 204).
>
> Do you think the study of small animals described in this chapter was relevant for the Year 5 children?
>
> Why were the children expected to catch the small animals in their school environment?
>
> As far as you can tell, given such a small sample of children's comments above, do you think the children regarded the technological task as: authentic, interesting, challenging, useful, important or meaningful?
>
> In terms of the three criteria (involvement, autonomy and relevance) significant from a social constructivist perspective, how would you critique the learning environment in the case study classroom?

SUMMARY

In this science and technology unit, the relationship between science and technology is a symbiotic one, with both components being valued equally. The unit, which was videotaped (Jane 1994) was designed to build on the children's previous experiences of plants in the school environment. It was evident that the children had ownership of their work because they could follow their interests and answer their own questions regarding the particular animals they found.

Through a symbiotic approach the children appreciated that there was a real purpose for making the technological products work effectively. Putting their products to the test, to see if they were useful devices to catch the animals and to contain them for the duration of the short-term study, enabled the children to reflect on their products' performance. In this way realistic appraisals were made, and the suitability of the materials evaluated.

Involvement in this unit developed the children's technological knowledge as they were able to suggest improvements, in terms of both the design of their products, and the type of material used. The approach adopted in this unit is best described as a symbiotic approach in which science and technology were valued equally. In the next chapter a process approach to technology is dealt with in detail.

REFERENCES

Gardner, P. (1990) 'The technology–science relationship: some curriculum implications', *Research in Science Education*, 20, pp. 124–33.

Gardner, P. (1993) 'Textbook representations of science–technology relationships', *Research in Science Education*, 23, pp. 85–94.

Jane, B.L. (1994) *Children Linking Science with Technology*. Video produced by Course Development Centre Video Production Unit, Deakin University, Burwood. Centre for Studies in Mathematics, Science and Environmental Education, Geelong, Victoria.

Jane, B.L. & Jobling, W.M. (1995) 'Children linking science with technology in the primary classroom', *Research in Science Education*, 25, (2), pp. 191–202.

Jones, A. & Carr, M. (1993) *Analysis of Student Capability*, (Working Paper No. 2 of the Learning in Technology Education Project) University of Waikato, New Zealand.

Keirl, S. (1998) 'Determining future directions for technology education: Can we?'. Paper presented at the conference of the Australian Council for Education through Technology (ACET), Melbourne, January.

Maruff, E. & Clarkson, P. (1988) *The Technology Studies Framework P-10*, Ministry of Education (Schools Division), Melbourne, Victoria.

McRobbie, C. & Tobin, K. (1997) 'A social constructivist perspective on learning environments', *International Journal of Science Education*, 19 (2), pp. 193–208.

Paige, K. (1998) 'Gear up gear down', in *Moving on ... All You Need to Know to Keep the Ball Rolling*, K. Paige (ed.) The Technology Teachers Association of South Australia Inc., Adelaide, pp. 15–19.

Stead, T. & Witt, C. (1998) 'Integrating technology'. Paper presented at the conference of the Australian Council for Education through Technology (ACET), Melbourne, January.

ACKNOWLEDGMENTS

Appreciation and thanks to Wendy Jobling and the year 5 children who participated in the case study research on which this chapter is based.

Chapter Six

The process approach to teaching technology

INTRODUCTION

The 'process approach' to teaching technology was briefly mentioned in Chapter 2 and now we look at it in more detail. This approach underpins national as well as many state and territory curriculum documents. The process approach is not only new to teaching, but new to many people's way of thinking about technology.

JOURNAL ENTRY 6.1 *Teaching technology education*

Revisit journal entry 1.6: What are your feelings about teaching technology? You were asked to consider how you might teach technology education and how you felt about this. Now that you have read Chapters 5 and 6 you will have some ideas about how to teach technology education.

How do your understandings compare with your previous notes?

It was shown in earlier chapters that many people associate technology with artefacts, with very few considering the technological process. The process approach to teaching technology explicitly deals with the process. However, how can early childhood and primary teachers begin to teach this process in ways which connect with children?

JOURNAL ENTRY 6.2 *Integrating technology education into the curriculum*

If you were out on practicum, how could you integrate technology education into the things that are already going on in the classroom or centre?

Record your ideas.

Since the release of the curriculum statements and profiles, we have heard teachers debate the benefits and limitations of outcomes-based education. Of particular interest has been the concern for how best to implement the eight key learning areas in ways which ensure connected and relevant learning experiences for children. As teachers begin to familiarise themselves with the curriculum statements and profiles (in whatever form they take in each state and territory) comments such as the following are heard (diary extract taken from a Year 2 teacher, Karina, in her initial introduction to technology education):

> 'Here is another key learning area to read about (technology statement and profile). Technology as a process is very new to me. How can I include technology in an already crowded curriculum in a way that will be effective and genuine for my children?' *(Karina)*

Technology has been met with great enthusiasm by those who have taken the challenge and embraced this key learning area. This same teacher analysed her program to see how to integrate a technological process into what she was already doing, by asking some key questions:

> 'Which areas of the curriculum am I teaching or planning to teach, which lend themselves to integration with technology?' *(Karina)*

This chapter outlines a range of ways that technology education can be incorporated into teachers' existing programs without compromising that program, the technological skills and knowledge advocated in the technology statement (Curriculum Corporation 1994) or the children and teachers themselves. The process approach is shown through a series of vignettes. Further ideas can be found in *I Can Make My Robot Dance. Developing Technology for 3–8 Year Olds* (Curriculum Corporation of Australia 1995).

TECHNOLOGY AS PROCESS

A process approach to technology education explicitly deals with *designing, making* and *appraising* (Curriculum Corporation 1994). This process is viewed within the context of materials, information and systems. In this section the process approach is discussed under the headings *of design, make and appraise; materials; information* and *systems*.

Design, make and appraise

In the design, make and appraise process, teachers assist children with using their plans or designs to make or do something. An example of this was shown in Chapter 5 where children were encouraged to draft up designs for a bug catcher. A further example may involve children planning their playground or outdoor area. The children follow their plan and organise their environment following their plan (which may include a series of games). The children then play in their new space (possibly playing the games as planned). This is the make or do phase of the process.

The next stage of the design, make and appraise process involves appraising or evaluating the product made or process undertaken. For example, in the previous chapter, the children evaluated their bug catcher based on how effective it was at catching and whether or not it could be used to house their bugs whilst they studied them. In the example of the children designing their own playground or outdoor area, the children can appraise their outdoor space against an established set of criteria determined by the children and the teacher (e.g. safety, movement between equipment, whether the games were fun or fair etc).

The design, make and appraise process is one strand of technology education as outlined by *Technology—A Curriculum Profile for Australian Schools* (Curriculum Corporation 1994). The other strands are:

- materials;
- information; and
- systems.

The designing, making and appraising strand features the following:

- investigate issues and situations;
- devise proposals and alternatives;
- communicate ideas and actions;
- produce processes and products;
- evaluate impacts and consequences (Curriculum Corporation 1994, p. 4).

Materials

Materials refers to the children coming to understand the properties of materials and how to work with them. Working with materials gives children opportunities to gain understandings and experience of a range of materials.

Materials students work with include fibres, clay, data, woods, film, ceramics, fabrics, soils, metals, plastics, plants, hormones and a variety of composites. The properties of materials can be used to create technological products and processes that meet particular needs and requirements (Curriculum Corporation 1994, p. 5).

Experience of working with materials ensures that children make informed decisions when selecting materials for particular projects. For example: 'Choosing materials requires careful consideration of advantages and limitations from technical, social and ecological viewpoints. These considerations influence decisions about the appropriate application of materials.'

'Students use a wide range of techniques to process, manipulate, transform and recycle materials, thereby creating particular structures, forms, effects and messages.' (Curriculum Corporation 1994, p. 5)

According to the Curriculum Corporation (1994, p. 5) when children work with materials they learn to:

- access the form, function, potential and suitability of material;
- select and use materials to achieve desired effects;
- use the physical, chemical and aesthetic properties of materials;
- use different types and combinations of materials;
- understand how the nature of materials affects outcomes;

- appreciate how the use of materials affects particular cultures and environments in Australia and in other parts of the world;
- create specific products and effects with materials;
- process, preserve and recycle materials;
- explore the past and present development of materials, and future possibilities;
- use materials safely and judiciously (Curriculum Corporation 1994, p. 5).

Even infants and toddlers work technologically as they learn to manipulate materials, as the following example illustrates.

Kay works in a nursery in a long day-care centre. She organises a technological environment by providing:

- playdough for children to manipulate, create and discuss;
- large blocks made from shoe boxes and covered in plastic, to build, climb over, bite, lift and drop (commentary by staff may focus on cause and effect or building work completed);
- a large tissue box covered in Contact, containing differently textured and coloured scarves for feeling, throwing through the air, piling as cushioning, looking through, playing games such as peek-a-boo, moving to music;
- sand and sand tray equipment for feeling, moving, changing (adding water), compressing and giving shape to by putting into containers;
- water and water trolley equipment for feeling, pouring, absorbing into sponges, giving shape to by pouring into containers and tubes, using as an energy source in waterwheels, displacing in containers, observing water and air pressure in tubes, submarines and bottles, as well as making comparisons between density and the shape of materials put into water and the upthrust of the water.

Figure 6.1 Silk scarves feel good and move well: the babies are learning about different materials through experiencing them

Infants actively explore the different materials, noting the properties and behaviour of them, forming general ideas about the materials, such as metals feel cold, some materials change shape like playdough, other materials are rigid, some materials are light (foam blocks) and others are heavy (bucket of sand). The adults in an infant's environment help label the experiences infants have through their contextualisation of commentary, such as: 'You can shape the playdough, Dylan.' Adults also set design challenges, for example: 'You have built a tower with the blocks. How high can you make it, Freya?'. Appraisal occurs as running commentary also 'Oh, it fell down. It was too high.'

How would you include the design or planning element of the design, make and appraise process with infants? Representational skills in older children demonstrate a higher order and more sophisticated understanding of how the world operates than with infants. Infants are absorbing vicariously through observation and directly through exploring with all their senses the materials in their environment. As toddlers, some are beginning to draw or model with dough in a representational manner. However, most are still more interested in seeing how paint can run down a page or merge with other colours. They are unlikely to represent anything on paper that they can use as a guide for making something. However, staff can plan with infants and toddlers. Through established routines, teachers can make lists or discuss verbally the order of things to come. Infants and toddlers can be invited to participate in these processes.

A range of examples of designing, making and appraising with materials follows in this chapter.

Information

The information strand involves the use of technological artefacts for the retrieval, storage and communication of information. Visual and sound images are often featured; for example, print, pictures, graphical representations and numbers.

According to the Curriculum Corporation (1994, p. 4) the techniques of gathering, sorting, storing, retrieving, transforming and communicating are all important processes in learning about information. Working with information gives children opportunities to:

- synthesise information in visual, aural, symbolic and electronic forms;
- edit, format and publish information in the form of texts, models, simulations and graphical representations;
- acquire and convey information to a variety of audiences through a variety of media;
- use and adapt hardware and software for managing information;
- create ways of organising and communicating information;
- understand the nature and uses of information;
- analyse, interpret and predict patterns and trends in information;
- assess the reliability and relevance of information;
- explore the social, cultural and political effects of information technology;
- gather, access, store retrieve, process and transform information;
- analyse and present information in ways that are gender inclusive and culturally inclusive;
- use data to synthesise and transform information.

Computers provide a powerful tool for achieving information storage, retrieval and communication. However, other technologies such as a video player or tape recorder are also suitable. Similarly, simple technologies such as pens, pencils and ochre paints also demonstrate other forms of information technology. An example of designing, making and appraising with information is shown below:

> Julie is a Year 3 teacher who worked together with her children to establish a radio station in her school. The school public address (PA) system was used to broadcast their program. However, before the children could go to air they needed to research their audience—what the other classes wanted to hear (nursery hour, rock and roll, news, talkback radio etc). They also had to research when was the best time to broadcast—during lunch, morning tea or first thing in the morning. To achieve their goals, the children needed to design a suitable program and then rehearse their broadcast (using a tape recorder). They also appraised their program before it when to air—live!

Figure 6.2   We have our own radio station in our classroom

Systems

The remaining strand detailed by the Curriculum Corporation (1994) is systems. Systems such as a roster, a bicycle, water drainage or recipes include separate parts or elements which are connected in a specific way to make the system work. According to the Curriculum Corporation (1994, pp. 5–6) working with systems gives children opportunities to:

- observe, dismantle construct, modify, operate and control simple and complex systems;
- investigate the performance of structures and mechanisms within systems;

- examine how systems are designed and applied to achieve specified outcomes;
- explore the forms, functions and performance of systems;
- explain how systems work and predict their cultural and environmental effects;
- use and develop organisational, electronic, mechanical, structural and information systems;
- understand how energy is used, converted and transferred in systems;
- examine inputs and outputs in whole systems and sub-systems within them;
- examine the appropriateness of particular systems in relation to different communities, gender groups and environmental circumstances;
- control and monitor the efficient and effective operation of systems;
- make, assemble, organise, manage and modify systems;
- evaluate the ethical implications of using different systems;
- examine the management and organisation of systems in particular cultures and environments in Australia and in different regions of the world.

An example of how learning about systems can occur is shown below.

Babies in child-care centres experience systems of care which ensure they feel safe and settled within their environment. Kay organises the environment in her nursery so that an overall routine is evident to the infants (greetings, exploring the environment, morning tea, exploring the environment, sleep, lunch, exploring the environment, sleep, afternoon tea, exploring the environment, departure), as well as providing them with individual routines that are familiar. For example, nappy changing times are a system of care which infants and carers use to not only physically care for infants, but further develop social and cognitive learning. A nappy changing routine may progress as follows:

Kay: 'Nappy changing time, Mary.'
Mary looks up and smiles.
Kay: (singing to the tune of Frère Jacques) 'Let's change your nappy, let's change your nappy, la, la, la ... (bends down and picks up Mary and moves to change table).

Each time Kay changes Mary's nappy, she begins her routine in this manner. As a result, the experience is consistent and predictable for Mary.

Another individual routine for nappy changing initiated by Nicholas follows:

Nicholas: (on change table) 'ch, ch, ch.'
Kay: 'You have chooks, don't you Nicholas?'
Nicholas nods.
Kay: 'Do you go and get the eggs?'
Nicholas: 'Ch, ch, ch, egg.'

Each time Kay changes Nicholas's nappy she talks to him about his chooks. This routine is a microsystem within the macrosystem of care in the centre for Nicholas.

continues...

102 TECHNOLOGY FOR CHILDREN

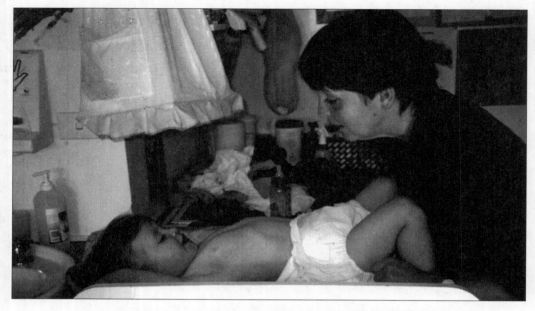

Figure 6.3   Babies experience a system when they participate in nappy changing routine

The infants in this centre not only feel that their world is predictable and safe, but that it is made up of a range of routines. These routines are the basis of a system. At such a young age, learning about systems involves experiencing what they are. Later, as children have gained more experience, they will be able to consider systems as discrete activities that they can appraise and change. Older children with adult assistance can actively plan, execute and evaluate their own systems.

TECHNOLOGY IN CONTEXT

In order for children to study designing, making and appraising information, systems and materials it must be introduced within a meaningful child-oriented context. When technology is considered as an integral part of what children already do and is situated in a meaningful context, the purpose becomes explicit for children. For example, when Karina looked at what she was already doing with her children in her classroom a great deal of technological activity was already taking place:

> A favourite activity for the children is making new inventions, making houses for pets and creating a variety of items from junk materials or construction kits. It won't be hard to tap into this existing activity and make the technological element of it more specific (*Karina*).

Karina organised her classroom so that the daily program provided many opportunities for self-directed and self-chosen activities through what she called *learning centres*. Her classroom contained a number of physically defined curriculum areas housing different resources and equipment. Children planned their activities for the session, carried out their plans and at the end of each day reflected on what they had achieved and learnt. At the end of each week individuals and small groups presented their work to the whole group.

In this environment it was very easy for Karina to focus on technology education. Karina was able to explicitly involve her children in thinking about the processes involved in technology education. As a result the children discovered that designing, making and appraising materials, information and systems is interrelated in nature. Karina and her children developed a number of key statements which reflected the technological process.

When we design, make and appraise (DMA) we:

1. Draw a design of what we plan to make and share it with the teacher.
2. Make the item, following the design.
3. Write about the item so that others can read about it.
4. Talk to the teacher about ways the item could be improved or changed, and include these ideas in the written explanation.
5. Display the item in the classroom or school together with the written explanation.

The children also found that they could start the technological process by making something first, then appraising what they had made, followed by drawing a design of the finished product! The children's chart below indicates this process.

When we make, appraise and design we:

1. Make the item.
2. Write about the item and then read about it.
3. Talk to the teacher about ways the item could be improved or changed, and include these ideas in the written explanation.
4. Make a design of the item or the appraised item.
5. Display the item and the design.

The process did not finish there for the children. They also had in their room the following chart:

We can also appraise, design and then make:

1. Play with the materials.
2. Appraise the materials and think about how they could be used.
3. Design the item.
4. Make the item.
5. Evaluate it.
6. Write about it and talk to the teacher.
7. Display the item.

As can be seen from the children's wall charts above, technology education is not a static or linear process. For these children it can begin as an idea from the children, from the materials themselves or from problems encountered—real life problems for the children such as how to organise a system for a class fire drill.

One of the important elements of Karina's program was the sharing sessions she organised for the children at the end of each week. This provided an opportunity in which children could share what they were doing, thus building common knowledge and cohesion across the group (and range of technological activities), as well as providing a forum in which children could ask for assistance.

'One group of children had been making a farm and wanted to put a pond in the yard. They had been experimenting all week with ways to make it work. Their comments during the session as we discussed their item were:

"We used plasticine to make a waterproof thing for the pond but it didn't work."

"Yeah, all the water leaked away overnight."

"Then we tried putting Gladwrap over the plasticine and that was better but still some of the water leaked away."

"Has anyone got any other ideas?"' (*Karina*)

Karina's diary (Fleer & Sukroo 1996) has given some insight into how early childhood and primary teachers are embracing this new key learning area. What she has demonstrated is how the process of designing, making and appraising with materials can be easily incorporated into teachers' existing programs. She has also shown how children can commence with either designing or making or playing with materials. Her work also illustrates how the appraisal element is an important feature of each stage of the process.

Technology education for Karina's children provided them with another way of thinking as they worked with materials, information and systems. Since the children organised themselves into small groups or acted independently, the charts displayed around the room (which had been jointly constructed by the children and the teacher) provided explicit steps they could use as they worked. The classroom context that had been established before technology had been introduced was maintained. However, the children now had a valuable thinking tool to use as they worked in the learning centres. Learning opportunities were broadened as a result of introducing technology education into this classroom.

INTRODUCING TECHNOLOGY TO YOUNG LEARNERS

Not all classrooms are organised into learning centres. In more traditionally organised classrooms greater effort is required to introduce technology education. Technology education is more than setting aside an hour in which the children are asked to build a bridge from newspapers with a view to it supporting a matchbox car! This activity is certainly technological, but the context is much more artificial than when a real human need arises in the classroom—such as organising a new lunch ordering and collecting system. The appraisal of this system is real—who would want the wrong lunch or one that is cold because it did not get to the classroom quickly enough?

Focusing on the physical environment

Human needs do not always arise in classrooms regularly enough for teachers to say they are implementing a technology program. Situations can be created which do not compromise teachers' programs—but rather enhance them. For example, children can be involved in designing, organising and using a new classroom layout. If the children are invited to design their own classroom layout a great deal of technological activity and learning can be generated.

Most young children will have well formed views about where they would like to sit in a classroom. Similarly, they will have ideas about the placement of equipment and materials around the room. Their plans may not always be safe, such as blocking a fire escape, or practical, such as placing the computer far from a power point. However, in using the room they will actively appraise these factors and re-design.

Children can be easily involved in re-designing parts or all of their school. This can be done inexpensively through a community project, where found or donated resources are used. Individuals or small groups can submit plans for consideration by a committee made up of adults and children. The stimulus could be a competition. The design brief for this project could involve all the children in the school. For example, interested children could observe how the existing school structures are being utilised. Individuals could be interviewed about their likes and dislikes. Potential problems relating to safety could be sought through interviewing the school nurse, organising a questionnaire relating to problem spots or talking to teachers on playground duty. Similarly, interested children could make enquiries about bullying or disproportionate use of the various areas in the school grounds. From this data, interested children could begin to develop a design brief. This design brief could also act as the selection criteria for the competition.

Focusing on existing routines

Whilst children can be actively involved in designing, organising and playing in their physical environment, they can also take charge of daily routines. For example, the children can brainstorm all the things they believe they do at school and begin to consider different ways the timetable can be arranged. Whilst this would need to be carefully orchestrated by the teacher through setting some non-negotiable perimeters regarding things that cannot be changed, such as lunch, these constraints would naturally fit within the brainstorm breaks. Whilst changing the whole timetable may seem too adventurous, there are routines within the day that can be modified by the children, such as news time.

There are special events which also take place in schools such as birthday parties, school fetes, athletics carnivals and assembly to name but a few. Children can also take part in organising these special routines. Bernadette, a Year 1 teacher, involved her children in preparing some of the food for their fundraising at their school tuckshop. In preparation for this the children brainstormed what would be involved. They decided to plan their menus, design advertising charts, jingles that could be played over the public address system, and invented chants that hawkers could use in the playground. At the conclusion of the day, the children went around and interviewed (using tape recorders) other children about what they had purchased, asking them to make any comments about how they had decided on their purchases. This information was summarised and shared with the whole group in an effort to appraise which advertising worked best!

As a follow-on from this work the children also started to critique advertisements in magazines, on TV and on the radio. In the process of doing this, a number of ethical dilemmas had to be worked through! Although the children were young in age, they discussed the gender stereotyping of many of the advertisements they saw on TV, heard on radio and read in fliers and magazines.

WORKING TECHNOLOGICALLY: WHERE ARE THE BOUNDARIES?

The technology education vignettes presented thus far indicate that technology can be broadly conceived. Within the process approach to technology teaching are many views on what form the content could take. Below are three more examples of technology education in practice. The first example (Clay in out-of-school-hours care) illustrates how technology education can easily move outside of the school or preschool walls. The second scenario (Where's that recipe?) presents an example of food technology. This is an area often not considered when people think about technology education, thus raising important gender issues in how technology education could or could not be constructed for children in schools. Finally, the third example (Old MacDonald had a farm) illustrates how broadly technology education can be conceived. Through an example of designing, performing and appraising a piece of music, the boundaries of what is and is not technology education are explored.

JOURNAL ENTRY 6.3   *Clay in out-of-school-hours care*

Context: The after-school-hours care program is situated in a small house on the school premises. The children who attend are free to engage in any of the experiences provided. Each room contains different activities. In the main area of the house can be found clay boards and clay tools. In the laundry is the kiln.

The children work together with the after-school-hours leader to construct a school banner made from clay tiles. The children decided upon the project, they planned the design of the banner and they each worked on producing a clay tile.

Is this technology? Record your views.

The manipulation of clay is an important part of learning about the properties of the material. When children know how clay can be worked (i.e. shaped, joined, dried and fired) they are in a good position to know how to make things from the material. With this knowledge and practical experience children can design artefacts which take into account the characteristics of the material. The act of learning about the properties of the material, designing and then making an artefact and appraising the process and the product can be considered as technology. Consider when it would not be technology.

> JOURNAL ENTRY 6.4 *Where's that recipe?*
>
> A group of six-year-old children are assembled in their classroom. They are working their way through a series of cookbooks. Each child is looking for recipes for their favourite food. They note the variations in the recipes (ingredients and preparation). On the following day the children collectively decide what their small groups will cook. After cooking and eating their food, they set about to invent their own recipes. The children write their own recipes, take them home to cook and bring back the product for their friends to sample.
>
> Is this technology? Record your views.

The children were engaged in technology education. They found recipes which they appraised. The initial cooking experience provided a common base (as well as a stimulus) from which to create their own recipes. The children's design was their recipe. Their appraisal took the form of trying out their recipe and tasting the product. With the clay example, the children needed to be familiar with the material. In the recipe example, the children needed to understand more about the process—the recipe and cooking procedures.

> JOURNAL ENTRY 6.5 *Old MacDonald had a farm*
>
> A group of preschool children are busy singing 'Old MacDonald had a farm'. Each child talks about what instruments they can use to simulate animal sounds. They use a cardboard cylinder to make 'cow' sounds and a triangle for the 'chicken' sounds. They progress to playing with the words in the song. Initially, they decide to change the name of the song from 'MacDonald' to 'Freya'. The song now begins ... 'Young little Freya had a farm'. The children use the tune to change other parts of the song—from a farm to a train 'Young little Freya had a train, chuff, chuff, chuff, chuff, chuff'. The children record their songs on paper and discuss which one they like best. Some of the children try to write notes to accompany the words.
>
> Is this technology? Record your views.
>
> If this is technology, is it designing, making and appraising with materials, information or systems?
>
> *continues...*

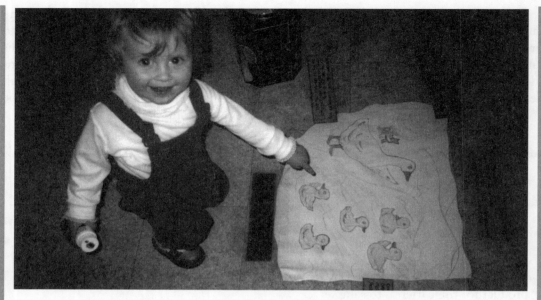

Figure 6.4 Five little ducks poster, contacted on to the floor in an infant's childcare room

This is also an example of technology. The children are designing, performing and appraising their song. They have used the tune of 'Old MacDonald had a farm' as the main part of their newly created song—this is the system. The words, actions and the musical instruments are the parts or elements which are used in their own creation of the song. This is a technological process. Although the product is not tangible, it is nonetheless a technological experience for the children.

Figure 6.5 Play dough work with babies is an important technological experience

There are many other ways that technology education can be introduced to very young children such as through children's literature, through nursery rhymes, and through art, music and drama. Technology education with its emphasis on designing, making and appraising with materials, systems and information extends children's learning opportunities. This framework provides another way of thinking for children as they engage in manipulating systems, information and materials. If we make the process explicit to children—as Karina did—then children will own and control the learning process.

SUMMARY

Through a series of vignettes, this chapter has detailed the *process approach* to technology education. The first section of the chapter provided an overview of each of the strands of technology as outlined in *Technology—A Curriculum Profile for Australian Schools*. The latter part of the chapter illustrated how each component works in practice. In Chapter 7 an *ecological approach* to technology teaching and learning is considered.

REFERENCES

Curriculum Corporation of Australia, (1994) *Technology—A Curriculum Profile for Australian Schools*, Curriculum Corporation, Victoria.

ACKNOWLEDGMENTS

The technological activities for infants detailed in this chapter were kindly provided by Kay Howell from Wattle Childcare Centre.

Vignettes originally published in full in the following publication were used to illustrate the 'process approach' to technology education: Fleer, M. and Sukroo, J. (1995) *I Can Make My Robot Dance. Technology for 3–8 Year Olds*, Curriculum Corporation of Australia, Victoria.

Ideas from the following article were incorporated into this chapter: Fleer, M. (1996) 'Working technologically in the early years', *Design and Education*, Special TEFA edition, February pp. 18–20.

Chapter Seven

An ecological approach: using technology appropriately

INTRODUCTION

One of the aims of this chapter is to broaden your perspective on global issues related to technology so that you can challenge children with appropriate technological tasks set in purposeful contexts.

NEW TWIST USING OLD TECHNOLOGY

Clockwork mobile phones and laptop computers will hit the Australian market as early as next year.

Clockwork radios, invented by Briton Trevor Baylis, are available now in Dick Smith stores for $149, and a clockwork torch will be launched within weeks.

The plastic FreePlay Radio is powered by a carbon, steel spring mechanism, the size of a fist, generating three volts.

Sixty turns of the radio's 'crank' will keep it running for about 30 minutes.

It is planned to miniaturise the system, but 150 spin-off projects are in the pipeline.

They range from home appliances to clockwork landmine detectors and wind-up global navigation systems which pick up satellite signals.

One of the world's oldest power sources seems destined for outer space, too. FreePlay spokesman Rory Stear believes clockwork equipment will soon become commonplace on space stations because it does away with the need for batteries.

'We have demonstrated we can do a laptop computer.' Mr Stear said.

'Anything that uses relatively little power is ideal. We will never do a hairdryer or toaster.'

But battery-free remote controllers, personal stereos, CD players, hand-held computer games, two-way radios, calculators and emergency lights are all on the way.

continues...

The Freeplay Radio and Personal Power Generation technology are the work of English inventor Trevor Baylis.

He reinvented the clock spring in 1992 while watching a BBC documentary about the difficulty of spreading AIDS information in poor and remote parts of Africa.

Mr Baylis came up with the idea of an electric motor powered by a hand-wound spring.

'Archimedes had his idea in the bath. I had mine for the clockwork radio when I was watching television,' he said.

Mr Baylis, 60, is a former army fitness trainer, movie stuntman and circus escape artist who had been working in the swimming pool industry.

He approached numerous British companies with his idea, but said: 'I was horrified by the number of rejection slips.

'I even had a letter from an august engineer telling me that the clockwork radio (which was merrily playing in my left hand) couldn't possibly work unless the spring weighed 100 pounds (45kg).'

Eventually, a frustrated Mr Baylis went public with his invention and within four days had struck a deal with accountant Christopher Staines who is now managing director of BayGen—the Baylis Generator Company.

South African Rory Stear joined BayGen and has built a factory employing disabled South Africans who can churn out up to a million units a year.

BayGen says Freeplay technology is making a difference in places such as Africa—controlling the spread of disease and educating the illiterate.

It is also proving popular with young greenies who see it as more environmentally friendly than batteries and electricity.

Apart from developing his BayGen business, Mr Baylis is campaigning for a Royal Academy to support struggling inventors.

By Murray Johnson

Figure 7.1 Sunday Herald Sun, *19 April 1998, p. 23*

JOURNAL ENTRY 7.1 *Environmentally friendly technology: clockwork radios*

Read the article 'New twist using old technology' which appeared in the *Herald Sun* newspaper on 19 April 1998 (see Figure 7.1). Several important issues are raised in this article which we would like you to think about and talk about with a partner.

1. What was the inventor Trevor Baylis doing when he had the idea for the clockwork radio?

    Recall the last time you had a really good idea. In your journal write down your good idea. Did you persist in following your idea through to its realisation? Why or why not?

continues...

When we ask children to think of ideas for their designs how much time should we allow?

Is it feasible to use children's good ideas as starting points for developing a design brief for a technological task?

2. Why do you think so many British companies would not develop the product invented by Trevor Baylis? How difficult is it in Australia for inventors to have their products developed?

It was necessary for the inventor to market his invention in order to receive the backing to develop it. Should we consider the marketing aspect when we teach technology to children?

3. How is the clockwork radio powered? What is a mechanism?

Why is the 'freeplay technology' proving to be popular with the greenies?

A TECHNOLOGICAL SOLUTION: CLOCKWORK RADIOS IN ATAKUA, GHANA

The following article appeared in an issue of *The Sunday Times* newspaper in England. The article contains the story of how John Knapton, a civil engineer and academic who lives in Newcastle, has helped the 520 Ekumfi people by giving them clockwork radios and extending their school. In the interview article John tells how he became chief (Figure 7.2).

'One of my former students, Kweski Andam, who was from the village of Ekumfi-Atakua, told me he was trying to raise money to tackle poverty there. Two years ago he invited me to come and have a look for myself. It was a horror story. The people were destitute. They had no water, they suffered from malnutrition ...; I decided that I had to do something about it.'

Knapton began raising money to buy the people clockwork radios. (The village has no electricity.) These enabled villagers to learn English and listen to flood warnings. 'They also heard that they should boil their drinking water. Consequently the mortality rate has been in decline,' says the big white chief.

Later, Knapton returned to the village with a team of students from Newcastle University to build an extension to the local school. So to show their appreciation for all his good works, Knapton was offered the honour of becoming chief.

Figure 7.2 *Extract from: 'Bright spark in the heart of darkness', The Sunday Times, 18 January 1998, p. 7.*
© Times Newspapers Limited 1998.

> **JOURNAL ENTRY 7.2** *Clockwork radios in Ghana*
>
> In an atlas look up the location of Atakua in southern Ghana, Africa.
> The introduction of clockwork radios has made a real difference to the quality of life of the Ekumfi people. In appreciation John Knapton, an Englishman, was made chief and what he says goes. This example shows the power that the distributors of technology can have. What is your reaction to this article?

Using technology appropriately involves a process of searching for an answer to a problem associated with people; for example, improving health problems caused by water supplies for a rural village community such as the Ekumfi. Information heard in English on the clockwork radios was an important step in the process of finding a solution to their drinking water problem.

In Chapter 1, in the section on appropriate technology, you considered the effects of the introduction of the snowmobile in Lapland (refer to Figure 1.6). This example illustrates the mismatch between the values of the designer to those of the consumer. You were asked to think of other technologies that have been transferred from one culture to another and to consider the results. Add the clockwork radio to your list and write down the effects of its introduction in the Ekumfi village in Africa. Don't forget to refer to the article in Figure 7.1 which tells of the factory where disabled South Africans are employed to make the radios.

Appropriate technology depends on *where* it is being used, *who* is using it and *what* it is being used for. 'Appropriate technologies' are right for the situation in which they are being used, and can be defined as 'the right tools and method for the job, or the right tools used with the right skills in the right context'.

> **JOURNAL ENTRY 7.3** *Predict the purpose of the mystery product from South Africa*
>
> In a small group consider the material from which the product shown in Figure 7.3 from South Africa was made. Write down each person's suggestion regarding what the product could be used for. Then come to a shared understanding of the purpose or function of the product and give the mystery product a name.
>
>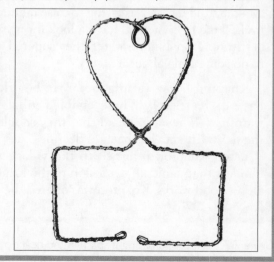
>
> Figure 7.3   Mystery product

Technology education has just been introduced into South Africa for Years 1 and 7, and will be phased into the other levels over the next six years (Shaw 1997). The example of technology you have been discussing in your small group (Figure 7.3) is actually a wire toilet roll holder used in homes in parts of South Africa.

In South Africa technology educators are designing modules for use across a continuum of life and school situations which address daily problems faced by the majority of South Africans (Harvest 1997). For example, the 'Water for Life' module addresses the issue that in South Africa there are 12 million people without access to purified tap water. In the case study of the implementation of the module, Year 2 children were required to design, make and evaluate a water filter in order to solve the problem of obtaining filtered water. The problem was contextualised differently in a developed school and a developing school. In the developing school it was found that the children's problem-solving skills needed encouragement in order to implement their own ideas. For many children the focused practical task—use of scissors—was the first time they had used a pair of scissors. Information resources available were only those provided by the teacher. The children were aware of the problems in their area, including the need for each house to have running water, but they had no idea about how this might be achieved.

APPROPRIATE TECHNOLOGY

Can technology experiences be designed so that parallel appropriate educational opportunities may be provided in developed and developing countries?

The concept of appropriate technology is not new. Twenty-five years ago Schumaker (1973) wrote about local economies developing 'appropriate intermediate technology' to serve local needs. Parkinson (1997) raises the issue that it is not a question of developing countries 'catching up' with the developed countries (which are adding to the global environmental problems), but rather finding an appropriate model for sustainability.

A new perspective, 'eco-development', has given rise to technology such as eco-technology or ecological engineering (Mitsch & Jorgensen 1989). In eco-development the shift is away from 'economic' fitness to one of 'ecological' fitness. For example, integrated agro-industrial ecosystems use environmentally responsible methods to produce food, energy and chemicals (Tiezze et al. 1991). Kinnear (1994) recognises that to achieve sustainability requires a rethinking of the way we assess, choose and use technology, and change is required locally, nationally and globally. Although action in local communities may influence institutional change, there must also be national policies and expenditures which encourage the innovation of sustainable technologies. Can this ecological approach be applied to technology education?

Several answers to this question can be found in Chapter 10, where the need to question our designed environment is recognised and explored. An ecodesign approach is suggested as a way of encouraging children to ask quality design questions of our environment. For journal entry 10.3 'Ecodesign teaching', you are asked to use a framework for stimulating quality design questions (Figure 10.1) to teach ecodesign principles. Examples (architects and building, Jelly Fruit Restaurant, all wrapped up) are given involving very young children which focus on environmental issues and ecological sustainment.

AN ECOLOGICAL APPROACH: APPROPRIATE TECHNOLOGY IN PRACTICE

An ecological approach involves developing appropriate technology. How can we decide what is appropriate technology for us? Remember from Chapters 1 and 3 that we should only evaluate the technological choices pertaining to our own culture, and focus on cultural similarities and shared needs.

The following list of 'criteria' can be used with children, first to explain the concept of appropriate technology, and then to evaluate whether the technology is appropriate.

- Is it what people need and want?
- Is it affordable?
- Is it made locally using local skills and materials?
- Does it generate income?
- Is it environmentally friendly?
- Is it controlled by the users?
- Is it culturally acceptable?
- Does it increase self-reliance?
- Does it use renewable sources of energy?

(Farrell 1997, p. 4)

Children can be encouraged to produce their own list of criteria for evaluating whether or not technologies are appropriate.

JOURNAL ENTRY 7.4 *Evaluating the appropriateness of technologies*

You can use the 'Is it an appropriate technology for us?' checklist shown in Figure 7.4 as a guide to evaluate your own design work and the work of other designers from a similar culture. Use the checklist to evaluate the appropriateness of the technologies listed by placing a tick or cross in each box. Add further examples of technology and check these against the criteria.

When completing the checklist, was it easy to decide whether the technology was appropriate? How did you rate the computer when applying the checklist criteria? When Aboriginal pre-service teachers from remote communities wrote down their prior views of what technology meant to them, the following responses were recorded.

Technology	Is it what people want and need?	Is it affordable?	Is it environmentally friendly?	Is it made locally?	Does it provide work and generate an income?
Bicycle					
Paper clip					
Talking doll					
Remote-controlled car					
Milk carton					
Gum boot					
Skateboard					
Knife and fork					
Personal stereo					
Hat					
Digital watch					
Telephone					
Computer					

(Adapted from Farrell 1997)

Figure 7.4 *'Is it an appropriate technology for us?'* checklist

> 'What technology means to me is data that is always being updated. You are always improving on your ideas which I think is technology. Cars, TV, sending satellites into space, radios, planes.' *(Eric)*
>
> 'To my knowledge technology is to do with everything involving the world: computers, buildings, cars etc and how we advance, how we use things.' *(Lynette)*
>
> 'A more challenging and convenient way of *improving* life. It also has implications and repercussions: job loss etc.' *(Donna)*
>
> 'Changes in today's world (e.g. computers, vehicles, speed etc.). Progress in life. Where man took the boomerang shape and made flight possible—aeroplanes.' *(Kaylene)*
>
> 'Technology is a form of improving things that are old and enhancing them. Computers, engineering, musical instruments, telephone, television, all communications.' *(Anne)*
>
> 'Technology implies inventions, machines, development, films. Technology forms a wide variety of sections of 'curriculum'. It describes ideas, products, processes and inventions required—or felt needed in contemporary to future society's 'needs' requirements: dishwasher, mobile phone.' *(Belinda)*
>
> 'To me technology is a study of things that are man-made; for example, cars, computers, also windmills and other things that work by wind.' *(Rosie)*
>
> 'Technology is the knowledge/skills to make and develop new ideas.' *(Judy)*

These comments are consistent with those reported in Chapter 1. It is interesting to note that the responses on the whole also detail 'high' technologies such as computers. Since the pre-service teachers were situated within a Western teaching context it is likely that these ideas will emerge. The effect of the 'social context' should never be underestimated when talking to people about their views on technology. An interesting follow up would be to ask the Aboriginal trainee teachers to mentally place themselves in their home communities and think about the everyday technologies they experience. In Chapter 3 we considered multiple world views and recognised the importance of teachers facilitating border crossing from one world view to another. Moreover it was argued that teachers should make overt the world view they are operating from. It would seem that these pre-service teachers are very capable of border crossing.

At the Centre for Appropriate Technology in Alice Springs the research organisation is actively involved in developing technological solutions for Aboriginal communities and isolated settlements. A two-way relationship has been developing, with researchers and teachers both applying and learning from Aboriginal problem solving in technology.

SCHOOL–COMMUNITY LINKS AS A WAY OF ENHANCING TECHNOLOGY EDUCATION

In the broader community, technology is often perceived in terms of mechanical or electrical devices, such as computers, fax machines and mobile phones. Recall from Chapter 3 that a high-tech approach forms one of de Vrie's categories to technology education. A design approach is advocated in the national project documents for the technology key learning area.

A recent survey was carried out with respect to the teaching and learning of science and technology in Australian primary schools. It was found that the majority of primary schools did not capitalise on resources available in their community to enhance their technology education programs. The Australian Science, Technology and Engineering Council's report (tabled in Parliament last year) recommends that primary schools make greater use of community resources in order to improve the teaching and learning of science and technology education (ASTEC 1997). The report encourages primary schools to foster links with community resources, such as the six science and technology (S&T) centres recently established throughout Victoria (at substantial cost to the state government).

It was the intention of the Directorate of School Education (DSE) that these S&T centres would form a network of interrelated centres of excellence and would provide support to all schools in the areas of professional development, curriculum materials and electronic networking. 'The centres are focal points and resources for the development of science education, technology education and the application of information technologies across the curriculum in all Victorian schools.' (Bell 1997)

> It follows that the view of technology that these S&T centres promote has the potential to influence how primary school teachers who visit them set about implementing technology in their own classrooms. For this reason we carried out a study of one centre, the Geelong Science and Technology Centre (GSAT). We found that a high-tech approach to technology education was predominant. One GSAT staff member perceived technology in a mechanical sense:
>
> > *I view technology as the use of any type of equipment in the classroom. It's not only computers but the use of cameras and videos and all those sorts of things ... In other words, it's using anything mechanical or technologically-orientated to complement your teaching in the classroom.*
>
> The study of GSAT focused on the perceptions of a small sample of primary school staff who had interacted with the centre in the first six months of its operation. The study found that GSAT was offering relevant professional development for staff and was setting up electronic networking facilities in schools. One advantage of GSAT being a private company is that it can provide schools with the hardware, software and expertise for on-line facilities. The 'Introduction to the Internet' program was popular for children's visits. The programs on offer clearly emphasise information technologies across the curriculum. This is not surprising given the current push by the Victorian government for learning technologies in all schools. Government funding is being made available to schools so that they can set up Internet and network links between schools. GSAT is responding to primary schools' needs by offering advice regarding the purchase of hardware and software so that learning technologies can be
>
> *continues...*

made available for student use. However, the data collected showed that programs did not adequately cover the design, make and evaluate approach advocated by the State technology document.

As expected there have been changes at GSAT since the study was carried out. One of the main changes is that the centre now has to be self-funding (as do the other S&T centres). The perceptions of the primary school users in our study revealed a tension between GSAT's commercial operations and its educational role. A further study would be beneficial in determining how the S&T centres have responded to the changed funding arrangements, and the repercussions in terms of their interactions with primary schools.

(Jane, Peel & Robottom 1998)

TOWARDS A FRAMEWORK FOR TEACHING USING AN ECOLOGICAL APPROACH

The significance of an ecological approach comes through in Paul Collins' (1995) book *God's Earth* in which he addresses the most important challenge facing humankind today: how to understand and redress the extensive destruction of the life-systems of planet Earth. 'I will always place the emphasis on the moral imperative of the *good of the planet* and of care for it, rather than on the *good life* for as many people as possible. Thomas Berry argues that "'the ethical framework that we need today must be provided by the larger context of the earth and the natural world"' (Collins 1995, p. 69).

> *The human community is subordinate to the ecological community. The ecological imperative is not derivative from human ethics. Human ethics is derivative from the ecological imperative. The basic ethical norm is the well-being of the comprehensive community, not the well-being of the human community. The earth is a single ethical system* (McDonagh, p. 514, cited in Collins 1995, p. 69).

Collins draws on the ideas of Heidegger who recognised that the technological process has been going on, in its modern form, since the 16th century and 'nature has become a giant gasoline station, and energy source for modern technology and industry' (Heidegger 1955, cited in Tropea 1987, p. 97).

As an ecological approach to teaching technology is an all-encompassing, organic approach with many interlinks and facets, there can be no universal framework. Different views of technology are represented as appropriate in different settings. The clockwork radio is a good example of how technology can be invented to make it suitable for use throughout the world whilst remaining environmentally friendly.

Throughout this book there are many examples of technological activities suitable for children which come under the banner of an ecological approach. For example, projects where the design briefs were initiated by the children themselves in order to address problems in their environment, such as the nesting boxes and bird feeders described in Chapter 4. Wherever possible the solutions generated should be sustainable, such as the temporary small animal enclosures made from recycled materials described in Chapter 5. In

Chapter 6 having children redesign their own classroom layout provides technological activity which can involve a process approach but which can also be considered an ecological approach depending on the materials used. Food technology, and designing, performing and appraising a song are also suggested as appropriate ways to introduce technology to children. These activities could also be categorised as an ecological approach to technology because they allow the children to follow their own interests (e.g. recipe for their favourite food) and relate to the environment (a song about Old MacDonald's farm).

SUMMARY

Children who are aware of appropriate technology can appreciate that technology reflects different cultures and values. Learning about technology in a global context can enable children to understand different approaches to developing and using technologies in a range of cultures. Finding out about new products, techniques and materials can help children evaluate the effects of technology on people and the environment. In the next chapter you will be provided with activities for you to do which will help you to position yourself so that you can develop your own approach to technology education.

REFERENCES

Australian Science, Technology and Engineering Council (ASTEC) (1997) *Science and Technology in Primary Schools*, Canberra.

Bell (1997) www.http//dse.vic.gov

Collins. P. (1995) *God's Earth. Religion as if Matter really Mattered*, Dove, Blackburn, Victoria.

Farrell, A. (1997) *Source to Sale: A technology education resource pack*, Intermediate Technology Development Group Ltd, Rugby.

Harvest, S. (1997) 'Water for life module—A case study to illustrate innovative curriculum developments within design and technology'. Conferencing proceedings of the International Primary Design and Technology Conference, Birmingham, June, vol. 2, pp. 62–5.

Jane, B. Peel, G. & Robottom, I. (1998) 'A case study of primary school use of the Geelong Science and Technology Centre (GSAT)', in *Enhancing Technology and Science Education through School Community Links*, B. Jane (ed.), Deakin University, Burwood.

Kinnear, A. (1994) 'A sustainable future: Living with technology. Society, choice and change, in *Framing Technology*, Leila Green & Roger Guinery (eds), Allen & Unwin, St Leonards, pp. 191–205.

McDonagh, S. (1994) 'Care for the Earth is a moral duty', *The Tablet*, 30 April, pp. 514–15.

Mitsch, W. & Jorgensen, S.E. (1989) *Ecological Engineering: An Introduction to Ecotechnology*, John Wiley & Sons, New York.

Parkinson, E. (1997) 'Some aspects of construction activity in design and technology education in Jamaica and the UK—a primary perspective'. Conferencing proceedings of the International Primary Design and Technology Conference, Birmingham, June, vol. 2, pp. 34–8.

Schumaker, E.F. (1973) *Small is beautiful. A Study of Economics as if People Mattered*, Blond and Briggs Ltd, London.

Shaw, P. (1997) 'Technology education in South Africa—the maternity ward'. Conferencing proceedings of the International Primary Design and Technology Conference, Birmingham, June, vol. 2, pp. 6–7.

Tiezze, E., Marchettini, N. & Ulgiati, S. (1991) 'Integrated agro-industrial ecosystems: an assessment of the sustainability of a cogenerative approach to food, energy and chemical production by photosynthesis', *Ecological Economics: The Science and Management of Sustainability*, Robert Constanza (ed.), Columbia University Press, New York, ch. 30, pp. 459–73.

Tropea, G. (1987) *Religion, Ideology and Heidegger's Concept of Falling*, Scholars Press, Atlanta, GA.

ACKNOWLEDGMENTS

Thank you to the pre-service teachers who provided their perceptions of technology.

The extract in Figure 7.2, 'Bright spark in the heart of darkness', *The Sunday Times*, 18 January 1998, is reproduced with permission of © Times Newspapers Limited, 1998.

Chapter Eight

Developing your own approach to technology education

INTRODUCTION
In the previous chapters you explored several different approaches to the teaching and learning of technology. In this chapter you further experience technology and reflect on these approaches, to help you develop your own approach to teaching technology. Teaching technology is like a technological activity, there are many solutions.

WHAT AM I?
I was purchased at a service station for the cost of $1.70.
My wrapper is made of purple-coloured foil.
I have a blue nose and a red mouth.
My name is Squish.
Unwrapping the foil reveals that part of me is edible.
Inside the chocolate there is a small plastic container.
Inside the container there are pieces of plastic which fit together.
There is a set of instructions which show how to fit the pieces together.
When complete they form an Australian animal.
The animal is a marsupial called a kangaroo.

JOURNAL ENTRY 8.1   *What am I?*

Write down your guess to the 'what am I?' question.
   In your tutorial you will be given a similar product to explore. What is the product's name?
   Crack the chocolate, open the plastic container and use the pieces to make the animal.
   Write down three things that you know about the animal you have made.
   When you have completed the task, refer to the steps given in the written instructions, and compare the way you went about making the animal with the instructions supplied.
   Check out its scientific name.
   How might you recycle the plastic container?

In the above activity you made a product (not of your own design) by trial and error until all the parts fitted together. People of all ages are collecting the Yowie animals, and in the process they are learning about native animals of Australia and New Zealand. The originator of this product idea was Geoff Pike who teamed up with Bryce Courtenay, author of the book *The Power of One*. Courtenay believes that through the Yowie, people will increase their awareness of native animals and how they interact with the environment. He is quoted in the cover story in the *Melbourne Weekly Times*, 1998 as having 'gone so far as to make a bold proclamation of the impact Yowie might have on the children at whom it is aimed: "I guarantee that in 10 to 15 years' time we are going to have some of the most sophisticated ecologists in the world, and it all began with Yowie" ' (Hunder 1998, p. 17).

Figure 8.1   Yowie adventures, Yowie creatures and Yowie chocolates!

Pre-service teachers had this to say about the Yowie.

'It depends on whether you are looking at it from a learning in science, or a nutrition point of view.' *(Chris)*

'We want them to learn about science, on the other hand the Yowie is encouraging them to consume, and it is advertising manipulation. There are cartoons and movies of shiney, furry creatures, who live in swamps.' *(Alan)*

'"Buy new Yowies" gives the impression that you can buy an ecosystem.' *(Mick)*

The 'What am I?' activity is an example of the *materialist science–technology* relationship where the technology (the making of a model) is the motivation which can lead children to a better understanding of the science concepts related to Australian and New Zealand native animals. The technology was introduced before the science concepts. This is one view of the approach which links technology with science. In Chapter 4 it was suggested that teachers should include technological experiences for children which cover the range of science–technology relationships, not just the commonly held view of technology as applied science.

Compare the various approaches to technology education put forward in this book by considering their similarities, their differences and the relationships between them.

> JOURNAL ENTRY 8.2 *Comparing different approaches to technology education*
>
> Refer to your journal notes on each of the approaches to teaching technology to children. Consider the advantages and disadvantages of several approaches and complete Table 8.1 by reflecting on your responses to journal entries 3.2. and 5.1. Mark with an asterisk the approach represented in your own state or territory documents.

Table 8.1 Comparing approaches to technology education

	Advantages	Disadvantages
Process approach		
Symbiotic approach		
Ecological approach		
Technology linked with science approach		

PUTTING IT ALL TOGETHER

As well as there being different approaches to teaching technology there are also different learning styles. In order to cater for the different ways children prefer to learn we should vary the way we teach (Malcolm 1998, p. 28). Figure 8.2 features personality types used when considering learner-centred logics for curriculum design.

> **THEORY BUILDERS:**
> Enjoy the power and logic
> of theories and generalisations
>
> **BIG PICTURE THINKERS:**
> Enjoy creating, relating and
> representing ideas in new ways
>
> **ORDERED THINKERS:**
> Are concerned about details and order;
> enjoy rules, set procedures, routines
>
> **FEELINGS THINKERS:**
> Are concerned about feelings,
> emotions, other people and lives;
> are intuitive, empathise

Figure 8.2 Learner-centred logics for curriculum design

We began many of the chapters in this book with a scenario or vignette of teachers' or children's comments in a relevant context. We chose to tell you stories about teachers and children and then invited you to reflect on the teaching and learning that occurred. What we did was to start in the 'feelings thinkers' corner and then move to the other corners shown in Figure 8.2. The Yowie activity above fits in the feelings thinkers' corner.

Cliff Malcolm contends that although in our lives we work with all four kinds of thinking we prefer one or two of them. The advantages of curriculum units which get into all four corners is that: 'they acknowledge and cater to the ways different learners like to think, and as fully functioning individuals all of us need to develop skills in all four thinking styles. It follows that learning how to think in these four ways should be outcomes of learning for all students.' Designing curriculum units and teaching 'depends on where the learners are, and where you want them to end up. That often means starting in the middle, even allowing different students to start in different places.' (Malcolm 1998, p. 44)

So that you can explore for yourself the kind of thinking various technological activities require children to engage in and hence develop, we encourage you to engage in several hands-on activities.

WHICH CORNER WERE WE IN?

In your tutorial group carry out the activities below and then for each activity decide which corner (theory builders, ordered thinkers, big picture thinkers, or feelings thinkers) you were in.

ACTIVITY 1: THE PROCESS MERRY-GO-ROUND

This activity develops awareness of a technological process and looks at ways to encourage its use in the classroom

Outcomes
You will be able to:

- Identify the components of a technological problem-solving process; and
- Understand different ways of using a technological process.

continues...

The starting point for planning a technological activity is generally the need or the problem to be solved. To address the need or to solve the problem requires a design brief which provides guidelines for the task. Consider the following historical example.

Canning as an example of technology arising from a need

- The *need*: In 1795 Napoleon (the French Emperor) needed to feed his army in winter.
- *Design brief*: A reward (12,000 francs) was offered for the invention of a practical method of food preservation.
- *Investigating, designing and producing*: Nicholas Appert, a French chef, preserved food by sealing it in glass bottles which he heated in boiling water.
- *Evaluating*: In 1806 the French navy successfully tried this food-preserving method with meat, vegetables, fruit and milk.
- *Improvements*: Tin cans were invented and in 1810 the first canning factory opened in London.

ACTIVITY TO DEVELOP AWARENESS OF A TECHNOLOGICAL PROCESS

Work in small groups to address the following design brief (Figure 8.3).

---

1. Investigate, design and construct to the prototype (model) stage; and
2. evaluate a mechanical aid to pick up litter.

The aid must have at least one moving part.

Figure 8.3 Design brief

Your group will be given a card which will assign you one of the following starting points: investigating, designing, producing or evaluating (Curriculum & Standards Framework, Victorian Board of Studies 1995). If you are the evaluating group you will be supplied with a range of products already made. Your group is to carry out the activity described on the given card in relation to the design brief.

Card 1 Investigating
Identify needs and opportunities, examine ethical, aesthetic, safety and health issues, consider past and present actions and future possibilities. Explore the appropriateness of a range of applications and their effect on the work of litter patrols.

> **Card 2 Designing**
> Generate plans and possible solutions to meet the design brief. Consider options, identify priorities and constraints, experiment with different ways to achieve outcomes, and calculate and predict consequences. Choose resources and appraise plans and actions. Use pencil design drawings, conventions and technical languages to explain design concepts and production processes. Use drawings to communicate ideas to group members.

> **Card 3 Producing**
> Work cooperatively to construct and trial a prototype. Select techniques and equipment, manage time and resources, and monitor and control quality in creating products and processes. Adapt ideas in response to constraints and difficulties.

> **Card 4 Evaluating**
> Develop and apply criteria to assess how well prototypes supplied meet specific needs. Measure and test performance, monitor human and environmental consequences, prepare oral and written reports, and make ethical and aesthetic judgments.

List of available materials:
Recycled paper, springs, dowel, glue, nails, masking tape, cardboard, plastic ice-cream containers, wire, screws.

Equipment supplied:
Pencils, rulers, erasers, hand saws, battery-operated drill, hammer, wire cutters, screwdriver.

Your group is to report back to the tutorial group about your involvement in the activity. Fit the pieces of a technological process together noting that there can be differing entry points. Sometimes a teacher may focus on the whole process, sometimes only one stage of it. Children usually toggle between the stages as Figure 8.4 shows.

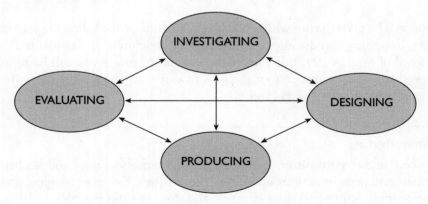

Figure 8.4   Techological stages

Conclusions

- There is no one correct solution to a design brief
- A technological process is not linear
- You did not use science or computers.

(Jane & Marshall 1997)

ACTIVITY 2: BEEF IT UP

A study of systems using the production of hamburgers as a model

Outcomes

You will be able to:

- Analyse the inputs, processes and outcomes that people can and need to control in a fast food producing system.
- Identify systems in a range of areas of human activity.

'Hamburgers on the run'

In your team of six organise yourselves so that you are able to prepare, cook and package six individual hamburgers (with at least three variations of contents for customer choice) within a time limit you specify.

Analysis of the system

- What inputs were there involved in the production of your hamburgers?
- What preparation of ingredients was required?
- What process did your team follow?
- What were the outputs from your system?
- What controls did you put in place and for what purposes?
- What management strategies did you use?
- What programming was involved in your system?
- How did each team's interpretation and implementation of the system vary?
- Identify the separate parts of your system and how they were connected to make the system work.

Look up your state or territory technology curriculum document and read the outcomes for the systems strand. In Table 8.2 below write down the learning outcomes (from the nationally produced document) for the hamburger activity in terms of each phase of a technological process.

Conclusions

- A system approach requires people to make decisions as a group.
- A systems view of production involves thoughtful consideration of production processes.
- A system approach requires an analysis of all the parts that make up the system.

Table 8.2 Learning outcomes

Level	Investigating	Devising	Producing	Evaluating
1				
2				
3				
4				

ACTIVITY 3: CRANK-STARTING TECHNOLOGY

Using technology as a focus in an integrated curriculum

Outcomes

You will be able to:

- Discuss how to start from a general theme and build in a technology component.
- Link a particular technology task to a theme.

1. Theme: China

In a small group in your tutorial brainstorm ideas for activities under each key learning area, for example:

English
Read about China and write a story about …

Health
Acupuncture and herbal medicines

LOTE
Language symbols
Mandarin

Mathematics
Money—overseas currency conversion rates
Population growth, density, graphs

Science
Life cycle of rice

continues…

Studies of society and environment
Land use and geography of the area—rice paddies
Map of China showing main cities
Religion

Arts
Culture
Opera—mask making

2. Technology: What technological activities can lead on from these areas?
Materials strand
In your small group design a model junk (a traditional sailing vessel) using recycled materials such as foam meat trays, wooden skewers, paper, foil etc. Make modifications to improve the model's performance after testing it. Use a stopwatch to time how long it takes for the model junk to travel a certain distance in a length of house guttering filled with water. A fan can be the wind source.

3. Link the technological activity to the science KLA
When testing each group's model junk in the tutorial link the performance to the science concepts relating to floating and sinking. For example, a junk with a broad, flatter-shaped hull will float better.

Conclusions

Ways to include technology in an integrated curriculum include:

- Begin with a general theme which is not technology-based. Build in a technology component by starting with a perceived need or problem and writing a design brief for a technological task.
- Select a technological task and put it into context by linking it to a general theme and link with science KLA by exploring relevant science concepts.
- Select a strand: materials, information or systems, identify a need or problem, and use it to write a design brief for a technological task.

(Jane & Marshall 1997)

JOURNAL ENTRY 8.3 *Thinking during a technological process approach*
Identify which corner you were in as you participated in each of the three activities above. What kind(s) of thinker does a process approach to technology promote?

In this chapter you are using the different thinking corners as a framework to assist you develop your own approach to technology.

ACHIEVING QUALITY IN TECHNOLOGY EDUCATION: WHICH APPROACH?

It is our hope that by engaging with this book you will gain the confidence and expertise to plan and implement effective technology programs in your school or early childhood centre, so that your children will produce quality products and achieve learning outcomes of a high standard. In Chapter 3 you were introduced to one way of achieving quality products in the materials strand by involving people in the community who have expertise in their chosen craft. The example described in that chapter was in the city of Birmingham where Asian women go into schools to assist with the teaching of embroidery, working with teachers and children in their technology programs in the textiles area.

In the United Kingdom inspectors visit schools and gauge how primary schools are implementing the national curriculum in D&T. Inspection findings found that:

> *good standards in Design and Technology Primary are mainly determined by the teachers' command of the subject, their confidence in teaching it and the influence that these have on the quality of the planning of the curriculum. Pupils' standards are high where the children:*
>
> - *regularly develop products to meet a simple specification;*
> - *make products using simple hand tools and a range of materials;*
> - *select tools and materials themselves thereby helping them to develop practical and organisational skills;*
> - *test and evaluate products, including those they have made themselves.*
>
> (OFSTED 1997, p. 4)

Inspection also found that pupils' attainment was limited because few teachers plan D&T curricula to include a balance of the following activities specified in the national curriculum:

- Designing and making assignments (DMAs).
- Focused practical tasks (FPTs).
- Investigation, disassembly and evaluation of artefacts (IDEAs).

There tends to be an emphasis on designing and making assignments, with few schools doing sufficient product analysis with existing products. The focused practical tasks enable the children to develop the skills required to achieve high standards. Where teachers provide children with opportunities to experience the full range of activities their attainment and satisfaction levels are high.

Stipulating these types of activities arose partly from concern that the DMA tasks being set for children were not encouraging progression in learning outcomes. It is intended that through structured tasks children will gain the technological knowledge and capability required to produce quality products.

In your tutorial group carry out the following focused practical tasks and IDEAs which are adapted from *The Coconut* (Farrell 1997).

IDEA 1
Split a whole coconut in two (put the milk into a container) and make an accurate drawing of the cross-section. Label your drawing showing the different sections of the coconut and describe some of the uses the parts might be put to. Discuss the uses of different parts of the tree.

IDEA 2
Compare the coconut milk with soya and cow's milk for taste, nutritional value, cost, preference and uses. Find out about the different ways in which milk is processed in Australia.

FPT 1
As a tutorial group, collect coconut products from home and friends, and set up a data base of your collection using a computer. Sort the collection into products which are similar (e.g. cosmetics in one group, food products in another etc).

FPT 2
Watch a TV advertisement for a typical shampoo (ideally one containing coconut). Discuss the images that manufacturers have selected to advertise this product. What messages are they trying to convey and why?

> JOURNAL ENTRY 8.4
>
> In your journal write down which thinking corner you were in as you participated in each of the four activities above.
>
> What kind(s) of thinker do the FPTs and IDEAs activities promote? Compare this response with the journal entry for 8.3.

ANOTHER LOOK AT AN ECOLOGICAL APPROACH

The coconut activities were originally devised as appropriate in the context of Kerala, a state in south-western India. In Malayalam, the state language, Kerala means 'land of coconuts' and thick groves of coconut trees are a main feature of the coastal belt. In Kerala the shell of the coconut is used to make spoons, ladles and dishes, and when polished makes beads and buttons. Appropriate technology has been defined as 'technology tailored to fit the psychosocial and biophysical context prevailing in a particular location and period' (Willoughby 1990, p. 15). It is designed to harmonise with nature not dominate it. The coconut was included here because coconuts and products made from them are readily available in Australia.

An appropriate 'focused practical task' was devised by Rochelle (a pre-service teacher in a school-based education unit) when she was teaching a small group of Year 1 and 2 children a unit on 'the senses'. For the sense of smell she decided that the children could make an environmentally friendly moisturiser. Rochelle sent the following letter home to parents so that they were aware of the contents of the product their child had made (Figure 8.5).

Dear Parent,

   As part of the Syndal Primary School Student teacher program _____ has been learning about the five senses and how in Science as well as the environment we need to be aware of all our senses.

As budding little scientists the children this week are learning about chemicals in the home. They learnt how scent is often added to chemicals to make them smell more appealing. E.g. Shampoo.

As part of their experimental experience the children have made moisturiser.

This letter is to inform you that it is safe for you to use. It contains Sorbelene, Glycerin, Water and a touch of _____.
The scent chosen for you by your child.

Enjoy!

Rochelle O'Keeffe
Student Teacher, Deakin University

Figure 8.5   Rochelle's letter to parents

Figure 8.6   John making moisturiser

The making of the moisturiser in this way as a gender-inclusive task, can help to avoid the stereotype of cosmetics being perceived as a predominantly female activity. On returning to the tutorial group Rochelle shared her experience of implementing this task with her fellow pre-service teachers. One male colleague chose to do the same 'make your own moisturiser' activity with his children the following week and it was equally successful. Chapter 11 takes up this important point in more depth.

At a higher year level children could make lip balm for dry lips from ingredients purchased in a natural cosmetic making kit or at health food shops or chemists. To make the lip balm, slowly melt half a teaspoon of almond oil and one tablespoon each of beeswax and coconut oil (melt in hot water) in a double boiler, stirring continuously. Pour into a glass jar and allow to set. Apply to lips. This is an oldie but a goodie.

Another oldie is goanna oil which Aboriginal people use as a moisturiser and they also cook with it. They also use beeswax in several ways. In Chapter 3 we considered the importance of taking into account different world views. During a tutorial session pre-service Aboriginal teachers discussed the topic of bees and honey and together with their tutor came up with the diagram shown in Figure 8.7 from which to plan a culturally appropriate curriculum.

BEES—product—honey

Aboriginal world view

Sugar bag—hunting and gathering
Eat the wild honey
Honeycomb
Tree stumps—black fly no sting

Western world view

Specialised roles
   queen bee (reproduces)
   drone
   guards

BEES worker—flowers—nectar—pollen—bee dance to communicate distance and direction of food

Uses of beeswax

Making didgeridoo
Shields
Protective coating

Candle making
Furniture polish

Figure 8.7 *Culturally appropriate curriculum*

An indigenous world view is very compatible with an ecological approach. You can readily see from this example that an ecological approach to technology can encompass links with science. It can include focused practical tasks as well as design, make and appraise tasks which involve a process approach.

> JOURNAL ENTRY 8.5 *Situating yourself*
>
> By reflecting on your experiences of technology and your journal entries, you should now be in a position to situate yourself in terms of your personal approach to teaching technology.
>
> Which approach do you prefer?
>
> Do we have to decide on only one approach, or can we have a combination of approaches?
>
> Should we vary the approach taken so that children experience (as you have done) a range of approaches?

In Part 3 you look at ways to plan for technological learning.

SUMMARY

In this chapter you were provided with a range of activities which were designed to encourage you to take a position. Journal entry 8.5 asks you to detail your personal approach to teaching technology.

REFERENCES

Earth First, *Natural Cosmetic Making kit*, Sandringham, Victoria.

Farrell, A. (1997) *Source to Sale. A technology education resource pack*, Intermediate Technology Development Group Ltd, Rugby.

Hunder, E. (1998) 'The yowie invasion', *The Melbourne Weekly*, April 7–13 1998, pp. 14–17.

Jane, B. & Marshall, A. (1997) 'Exploring technology'. Unpublished workshop materials for the Technology Key Learning Area, Victoria.

Malcolm, C. (1998) 'Thoughts from South Africa—begin at the middle', *LABTALK* , vol. 41 (1), February, pp. 28–9.

OFSTED (1997) 'Inspection findings in design and technology—1995-96. *DATANEWS*, No. 5, April, pp. 4–7. Victorian Board of Studies (1995) *Technology Curriculum and Standards Framework*, Carlton.

Willoughby, K. (1990) *Technology Choice: A Critique of the Appropriate Technology Movement*, Westview Press, Boulder.

ACKNOWLEDGMENT

Thank you to Rochelle O'Keeffe for her activity and photographs.

Part 3
Planning for technological learning

Chapter Nine

Finding out children's technological capabilities

INTRODUCTION

In early childhood settings the word 'technology' conjures up thoughts of building blocks, Lego, Mobilo and many other types of construction. The teacher structures the environment, provides the materials and then retires from the scene so the children can play freely. We then claim that we are preparing and exposing children to experiences, ideas and products that are found in the world of technology.

JOURNAL ENTRY 9.1   *Children's technological capability*

How would you interpret the following transcripts taken from a series of interviews between four and five year olds and Marita, their preschool teacher?

Lauren: 'I'm going to build a house with a kitty cat out the front.'
Marita: 'What do you need?'
Lauren: 'You have to get some materials and tiles … the things you need for houses.'
Marita: 'What is the first thing you need to do?'
Lauren: 'Draw up your ideas and plan what you are going to do with your house.'

Marita: 'How do you make a pirate ship?'
James: 'Design it.'
Marita: 'How?'
James: 'Draw how I will build it.'
Marita: 'What goes in a design?'
James: 'A mast … some space where they can sleep at night … a wheel of a ship and an anchor. Then I will build it.'
Marita: 'Why the design?'
James: 'So we don't forget!'

Jessica: 'You can make a kite.'
Marita: 'How do I know what the kite will look like?'
Jessica: 'You just make a design and copy it.'

Record your response in your journal.

It was noted in Chapter 2 that children have had a range of experience with technological products and processes before commencing school or preschool. The vignette described above gives some insight into what children are capable of doing and challenges educators to carefully plan experiences which will extend children's capabilities. Contrast that vignette with the following teaching example organised for a group of seven year olds.

Teacher: 'I would like you to use the newspapers and the sticky tape to build a structure to support a matchbox car. There is a box of matchbox cars over here. Each group can choose one. Any questions? (three second pause). OK off you go.'

- Is the teacher able to determine the children's prior technological experience or their technological interests?
- Will this experience build upon children's technological capabilities?

This chapter illustrates how teachers can organise learning environments which not only determine children's technological capabilities but build upon and challenge learners.

ORGANISING LEARNING ENVIRONMENTS

In organising technological learning environments for young children, educators need to ask themselves two fundamental questions:

- How can educators find out what young children already know?
- How can this process be a part of the teaching–learning sequence?

The three teaching examples shown below ('The fairy room'. 'The lonely creature' and 'The imagination space') illustrate how to build into a teaching–learning sequence the ability to assess children's technological capabilities in a meaningful way. The examples also demonstrate how to build upon the children's technological capabilities so that learning opportunities are maximised by children.

The first example illustrates how technological learning can be organised in a preschool environment. Marita, the teacher, removes all the items from the home corner area prior to the children returning to preschool after the mid-year school holidays. The children are encouraged to collectively use the space in whatever way they wish—hence setting up a technological challenge for the children as they make decisions about what to do, how to plan their space, find ways to collect materials and actively construct their new space. The children decide to construct fairy homes.

The second scenario focuses on planning technological learning for three and four year olds attending a childcare centre. Wendy, the teacher, begins her teaching–learning sequence by telling the children a story about a lonely creature she found in the garden. The creature is hidden inside her basket and won't come out because it is frightened and wants a friend. The children are invited to construct a friend for the creature and later a house. Wendy encourages the children to use the collage and box construction materials to make a friend. She also announces that should the children need other things, they are to tell her and she will put them on the list—ready to purchase that afternoon for use the following day. As a result, the children are encouraged to plan what they need.

In the third example, Shirley and Mary, who team teach a group of Year 1 children (seven year olds), create a space in the classroom in which the children can collectively make a structure. The children decide they wish to use the construction materials available to make a puppet theatre.

The three open-ended examples are shown below in detail. As you read, note:

- how the children decide on the design brief;
- that the open-ended approach allows the teacher to map individual children's capabilities;
- how the technological task is meaningful and relevant to the children, thus stimulating children to work to their potential;
- the way in which the teacher is free to move about and make assessments on individual children's capabilities;
- the limited use of materials and minimal preparation;
- the technological experience is designed to extend over time (i.e. two weeks); and
- other key learning areas such as literacy and mathematics that are also featured in the units.

 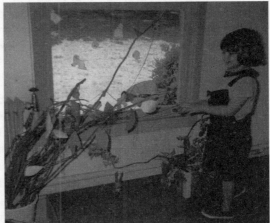

Figure 9.1   The Fairy Room

THE FAIRY ROOM (preschool)

In this unit I concentrated on finding out about children's technological knowledge, skills and language.

Creating the Fairy Room
After the holiday break the children came back to their classroom. Immediately they noticed a difference.

Elliot: 'It's (the furniture) all gone into there (points to main classroom). All the home corner is gone. The drawers were here and now they are out there.'
Lauren: 'It's empty … it's called nothing … not a thing in it.'

continues…

The children discussed what they could do in the room

Ollie: 'I think we could have fairies living in this room.'
James: 'And then we could have a garden for them.'
Helen: 'We would need a house for them.'

Hence the Fairy Room was invented.

Why did I plan the concept of a room that the children could use in what ever way they wished?

I believed it would be an interesting, unusual and effective way of encouraging children to introduce technology to their peers. It would be a setting that could provide children with practical activities to test their abilities in planning and designing. It definitely invited every child to participate in a co-operative group project and ultimately offered the ideal opportunity for every child to be a successful technological person. By having control of their own project the children would have to identify plans, make decisions, be involved in choices, take risks, solve problems as well as investigate issues, devise proposals, produce products, continually re-assess and evaluate their work. WOW !!! Just imagine what this will do for their understanding and progress in the technology area, let alone their self-esteem and belief in themselves and their positive contribution to the class.

What was my role in relation to the Fairy Room?

My role was that of a non-active participant. Through home visits and individual interviews I was so aware of the children's understanding and enlightened attitude towards planning and their ability to organise and be in control of their own lives. So I became the adult who would audio tape and transcribe their conversations, who scribed on the whiteboard and labelled their designs and documents. I listened to verbal reports and watched them interpret and solve problems. I took photographs daily and took notes on their research skills. I observed and followed their designs and work from a distance and became one of the many guests invited into the Fairy Room. What a powerful opportunity and learning tool this was for me in being able to record and document young children's involvement in the process of designing, making and appraising with materials.

(Marita)

The Fairy Room generated a high level of interest and excitement for the children. The notion of a Fairy Room quickly expanded to building a fairy house. As the children discussed their ideas, four separate groups became established in the classroom, each with their own beliefs, understandings and expectations of the Fairy Room.

Group One:
'Fairies are only tiny, so a fairy house could only be as big as a toothpaste box.'

Group Two:
'Fairies don't live inside, they live outside in the garden.'

Group Three:
'Fairies are not real, they are only in your imagination.'

Group Four:
'This is the fairy house we are going to build.'

The first group organised a construction table of boxes, tape, paper, wool, twigs etc. They set about making lots of fairy houses on a miniature scale, complete with windows, doors and some household furniture. When completed, the children placed their fairy houses up on the window sills: 'So the fairies can have light and see the sky.' (Jarred)

The second group stuck to their beliefs that fairies live outdoors—under mushrooms and in the flowers. This led the children to prepare a village site under a large pine tree. The children made paths, roads, hills, mountains, roundabouts, wooden structures and even garden beds. This was the site for the fairy houses. The children used toy trucks, spades, bark, grass and dirt to help them with their constructions.

This group worked on their fairy village each outdoor time. Andrew drew a map of how to find the fairy house in the outdoor area. The children's belief of the outdoor homes was strengthened by finding fairy mushrooms (red and white) growing under the trees after heavy rains.

The third group chose not to participate in the project establishing a fairy room or building a fairy house because fairies are not real. Hence their was no interest in participating. Their rights were respected.

The fourth group was comprised of children who had drawn up plans and designs for a large-scale fairy house. They pooled and shared their ideas:

'It must have a small door with a little door handle.' (*Alexa*)

'It must have a corridor that takes you to the bedroom.' (*Helen*)

Plans included minor details such as: 'An alarm clock to wake up the fairies' (*Ollie*) to major structural details of footings: 'We need to put cement here first, so that the house can stand on it' (*Tim*) and 'The rooms need to be square' (using a set square). (*Alexa*)

The children drew on their experiences and understanding of building requirements for house designs. The children's plans included roofs, floors, walls, doors, windows, door handles, stairs, corridors and carpets. The rooms were labelled as kitchen, toilet, bedroom and bathroom. It was interesting to note that some of the children's designs included indoor and outdoor features: 'The flowers grow in the soil outside' (Ollie).

Once armed with plans and designs the children constructed a house using timber blocks, tables, plastic shelves and other suitable materials they could carry into the Fairy Room. A real team effort!

Through the Fairy Room Marita was able to take children's thoughts, experiences and current knowledge and use play and imagination as a tool to provide the children with technological learning. The Fairy Room provided the magic blend of play, imagination and technology, thus creating an environment in which the teacher could more accurately assess children's technological capabilities (adapted from Fleer, Corra & Newman 1996).

The Lonely Creature

Wendy's task was to set up an environment in which her children felt motivated to design, make something and appraise it. In particular, she was interested in whether or not the children made an effort to plan their work prior to making it. Wendy decided to use box construction as the medium as it was readily available and the children were familiar with using the construction trolley independently

Wendy brought in a picnic basket and told the children a story about a creature that she had found in the dark cold night, sobbing beneath a bush in her garden! The creature was lonely and needed a friend. Since the basket concealed the creature—and it was too frightened to come out—the children had to imagine what it looked like and think about what sort of friend they could make to keep it company.

THE LONELY CREATURE (childcare)

'The lonely creature is too frightened to come out of its basket. Can you make some friendly creatures to keep it company?

The materials available include: collage materials; boxes, cylinders, popsticks, match sticks, woodwork glue, masking tape, scissors and staples.

You should make a list of anything extra that is needed or to be purchased, for the next day.'

Day One:
The children were told the story whilst I held a basket covered in a blanket on my lap, supposedly concealing the lonely creature. The children's faces were filled with empathy and most of them were very eager to construct a creature to keep the lonely one company.

Day Two:
The activity was repeated. However, this time, I asked each child to describe their creature and what they needed before we went to build them. Later that afternoon, I talked to each child, telling them that the next day, I was going to bring in whatever equipment they needed to make a house for the creature. They each listed the equipment which they would need and drew a picture of what they thought the house would look like.

Day Three:
The children were presented with the requirements which they had listed; for example, one child asked for six bricks and when he arrived, there were six bricks on his table. We talked a lot about how we could join the materials together—what techniques worked well and which did not, when reinforcement was needed (for example, extra tape over glued boxes). We also explored safe ways of using scissors and staplers when cutting cylinders and boxes.

(Wendy)

The open-ended experience that Wendy had organised allowed her to record how individual children were planning, making and appraising with materials. As a result, Wendy gained a great deal of understanding about how to go about planning future technological learning experiences for her children.

My feelings about these days of watching and wondering how children think resulted in the following conclusions:

Young children have the ability to plan their activities. An example of this was Teddy's house. He drew a picture of his house prior to building it, listed the precise materials required and then built an identical replica of his planned drawing. (*Wendy*)

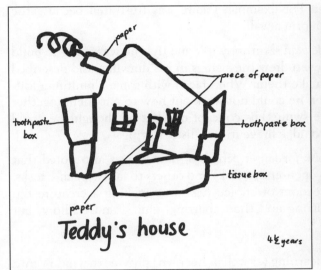 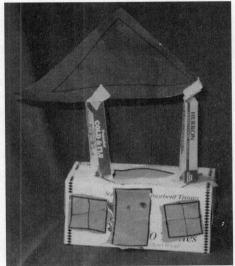

Figure 9.2   Teddy's design and model

Teddy's design can be seen in Figure 9.2.

Technology—A Curriculum Profile for Australian Schools (Curriculum Corporation 1994) is useful in analysing Teddy's design capabilities. Teddy is working at Level 1: generates ideas for own designs using trial and error, simple models and drawings. Wendy comments on the children's design capabilities:

The children's ideas are always flexible.

The children are influenced by the materials on the construction table. Many examples of this were seen as the children really wanted to use materials which suggested in themselves a creature of some sort. James found a toilet roll which had been previously covered with cotton wool and although he had told me he was planning to make a kookaburra, he decided to use the roll to make a sheep.

The children are influenced by ideas incorporated by their peers. On Day Three, Angela had requested white paint in order to paint her house. The children at her table thought the white paint was an exciting alternative and proceeded to paint as much as they possibly could in the allocated time!

Children developing a sense of their limitations regarding the physical skills required. An example of children being limited by their inability to cut, stick and paste was demonstrated by Daniel. By the end of Day One, he had carefully laid two sticks for the legs, two pipe cleaners for the head, two straws for the wings, two patty pans for the eyes, and cotton balls to 'hold the body up'. He could not go any further because he didn't know how to attach them all. Similarly he could not attach the pieces of his house.

Children developing a sense of the limitations regarding technical knowledge. Many children planned great things! When it came to making them, however, they did not have the knowledge to put things together. For example, Claire was frustrated because the paper would not remain upright to form a wall.

A developing ability to form a realistic plan. Anthony told me that he was going to build some swings for the creature. He even drew the swings in his drawing, and described the metal rods he would need to make them. When faced with some aluminium foil-covered paddle pop sticks, however, he could not work out how to put them together and told me he couldn't make the swing. His concept of what he thought he could achieve was different to what he could achieve in a realistic setting.

A developing ability to remain on task for longer periods of time. It was also noted that some children responded better to open-ended tasks and others to fairly specific tasks. For example, Claire found the freedom of choice regarding making a creature too great. She spent a lot of time cutting and then chatting, gluing and fiddling, but without much of a result.

Wendy's description of her children's learning (as well as her own) provides a window into children's thinking processes in terms of designing, making and appraising with materials. The open-ended experience allowed children to bring to the task—making a creature—their current planning, making and appraising skills with box construction materials. Wendy could record from this an individual profile for each child, and plan further technological experiences which would build upon individual children's existing skills and knowledge. Although the document *Technology—a curriculum profile for Australian schools* (Curriculum Corporation 1994)—was useful in analysing the first stage of the children's design capabilities, Wendy's commentary provides assessment details which will be useful for further planning. For example, Daniel (who created a detailed design) was unable to adhere the legs onto the body of his creature. His design did not recognise the practical constraints inherent in joining materials (Devising 2.2: generates designs that recognise some practical constraints using drawings, models and, where necessary, introducing some technical terms). In another example, Claire was unable to generate a design for her creature. Her attention was on the range of materials. She needed assistance with recording her ideas on paper before beginning to produce a creature. Devising Level 1.2 specifies that children can: generate ideas for own designs using trial and error, simple models and drawings (Curriculum Corporation 1994). Claire was clearly not at this level.

Wendy talked to the children about their designs and their constructions to see how they felt about them (Level 1.4 Evaluating: describes feelings about own design ideas, products and processes). The work samples shown in this unit and the scribed comments (shown below) demonstrate how even very young children are able to clearly state what they think about their work.

'I would like to change it (crocodile) into an elephant!' (*Alicia*)

(Giraffe) 'I would make a longer neck, a longer tail. I don't like anything.' (*Teddy*)

(Hot air balloon) 'I'd make an animal. I don't know yet. I liked it because I could play with it in my bath!' (*Sean*)

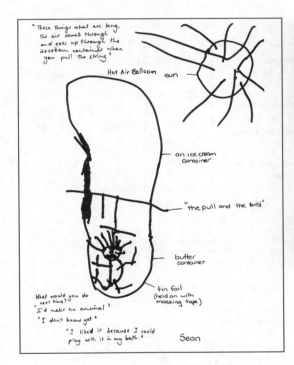
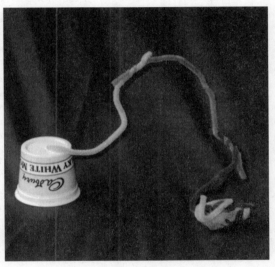

Figure 9.3   Sean's design and model of a hot air balloon

(*Source*: Adapted from Fleer & Sukroo 1995)

In this unit ('The lonely creature') assessment occurred within the framework of existing materials (box construction and collage materials) and processes (collage area of the childcare centre). The stimulus for encouraging group activity was the storytelling. The deliberate use of an imaginary creature and the use of a 'shopping list' to encourage planning were also included in this unit. Collectively, the approach adopted allowed a more accurate profiling of children's technological capabilities. The overall scope and sequence of technological learning detailed in the profile provided guidance for this task. However, the fine gradation of assessment necessary for future planning was done by Wendy in her analysis of how children planned, designed and produced with materials over the entire production process. The complete technological process for each child needed to be considered when making assessments of children's technological capabilities. 'The lonely creature' unit allowed assessment over time and over a number of processes (e.g. planning and making the creature, making a home for the lonely creature).

The Imagination Space

The Imagination Space created by Mary and Shirley provides a further example of how teachers can profile children's technological capabilities. This example is from the school context. Mary and Shirley outline how they set up the Imagination Space and the learning outcomes generated for them as teachers.

THE IMAGINATION SPACE (first year of school)

We wanted to give the children a space in their classroom where they could do their own thing, a place we had not organised for them. We called this the Imagination Space. We have done a lot of talking with the children as a group. We talked about 'What is imagination?'.

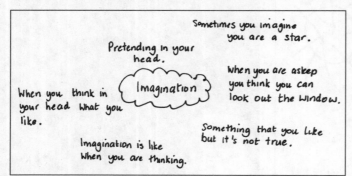

We thought that was a basic idea we had to establish first. We were amazed at some of the concepts the children had. (Figure 9.4 shows the group concept map of 'imagination'.)

Figure 9.4 Group concept map of 'Imagination'

From there we started to explore the idea of planning from the ideas in their heads. What did it mean? Again we were delighted at the concepts the children had—drawing a picture, writing or making a model of their ideas. We needed to spend quite a lot of time on the planning process before we went on to the Imagination Space. (Figure 9.5 shows the 'planning' map.)

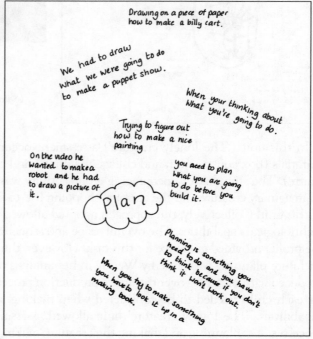

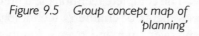

Figure 9.5 Group concept map of 'planning'

We sat down together as a group and decided to plan 'What we could do in the Imagination Space'. We brainstormed a whole host of ideas; for example, airport, police station, farm. A common idea was that of a puppet theatre.

The children decided they wanted to use the box of quadro that was in the school to help them with constructing the puppet theatre (Figure 9.6).

From this whole process we have gained a great deal of information about what children know about planning, about how to build effectively and what they can do.

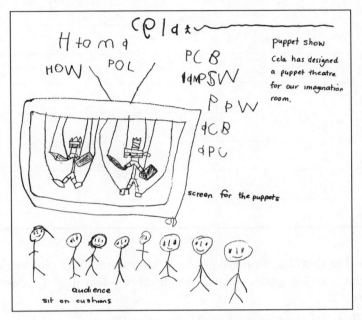

Figure 9.6   Cela's design of a puppet theatre

We now know a lot about the children's technological capabilities. It was because the children controlled the activity, that we felt we were able to assess their technological capabilities accurately. For example, Cela's design demonstrated that she already understood a great deal about puppet construction. She included detail in the drawing of how the strings were attached to the puppet limbs. However, she was careful to obscure the top string frame and hands. This indicated that Cela was able to generate designs that recognise some practical constraints using drawings (Devising, Level 2: Technology—A Curriculum Profile for Australian Schools).

(Mary and Shirley)

With the information Mary and Shirley gained about their children, they were able to make informed decisions about their future planning in technology education.

ASSESSING CHILDREN'S ACHIEVEMENT

In the past educators have written behavioural objectives, organised learning experiences to meet these and then assessed children's progress in relation to the behavioural objectives. For example:

> By the end of the lesson most children will be able to adhere a cylinder-shaped object to a rectangular-shaped object.

However, with an outcomes-based approach to the planning, learning and assessing cycle, educators have had a new way of thinking about assessment. Since the release of the Profiles assessment document (*Technology—A Curriculum Profile for Australian Schools* 1994) educators have been encouraged to map what children are able to do. This information is used to profile individuals, thus forming the basis for future planning (see Table 9.1).

Table 9.1 Profile of Cela's technological capability in relation to materials

LEVEL	Investigating	Devising	Producing	Evaluating
1	*	*	*	*
2		*		
3				
4				
5				
6				
7				
8				

The three examples presented in this chapter highlight how the assessment of children's technological capabilities can be integrated into the teaching–learning process. Assessment of children can be:

- designed around open-ended units such as the Lonely Creature, Imagination Space or Fairy Room;
- conducted within the full context of the learning experience;
- holistic rather than compartmentalised;
- conducted over time and over production processes; and
- sufficiently meaningful for children to exhibit their full potential.

Assessment is about finding out what children are capable of doing. Assessment is a tool for planning. Siraj-Blatchford (1997, pp. 68–9) suggests that when teachers use assessment documents (such as Profiles) they should:

1. identify the stage the individual or group is currently working at at any specific time (e.g. are they still designing, making or have they moved on to evaluating?);
2. identify the level at which they are working. Also, identify those aspects of the program that have been covered at that level;
3. provide guidance as to the support required, and guidance for the teacher in directing their pupils towards the higher levels within the stage they are currently working;
4. provide guidance for teachers in supporting and encouraging the pupils in moving to the next stage of the process.

SUMMARY

Educators need to focus on identifying the needs of individuals, rather than on the differences between individuals. It is the progression that is important and not the race to a predetermined end point. The challenge for educators is devising alternative assessment techniques to the testing of all children at the same time. The three examples in this chapter provide a beginning point to devising appropriate ways of determining children's technological capabilities.

REFERENCES

Curriculum Corporation of Australia (1994) *Technology—A Curriculum Profile for Australian Schools*, Curriculum Corporation, Victoria.

Fleer, M., Corra, M. & Newman, W. (1996) 'Playing around with technology: children creating multiple learning pathways', in *Play Through the Profiles: Profiles Through Play*, M. Fleer (ed.), Australian Early Childhood Association, Canberra, pp. 47–56.

Fleer, M. & Sukroo, J. (1995) *I Can Make My Robot Dance. Technology for 3–8 year olds*, Curriculum Corporation, Victoria.

Siraj-Blatchford, J. (1997) *Learning Technology, Science and Social Justice: An Integrated Approach for 3–13 year olds*, Education Now Publishing Co-operative, Nottingham.

ACKNOWLEDGMENTS

Marita Corra, Wendy Newman, Mary Crix and Shirley Gollings provided the teaching vignettes in this chapter. Their contributions are highly valued and much appreciated.

Chapter Ten

A question of design

INTRODUCTION

Figure 10.1 Why is liquorice twisted?

Why is liquorice twisted?
Why does my chair have a curved back?
Why does our chair at home not have arm rests?
Why is there netting over the roast pork?

These technological questions were generated by a three-year-old child. A child who genuinely wanted to know why her constructed environment looked the way it did. Why *is* liquorice twisted? We take for granted liquorice and never question its design—like many other things in our environment. However, children notice and want to know why things look the way they do.

> **JOURNAL ENTRY 10.1** *Asking design questions*
>
> Refer back to journal entry: 1.4 'Our technological environment'. Examine what you have listed.
>
> Have you ever questioned the way your environment is organised or constructed?
>
> Did you ever think about why the chairs were designed as they are?
>
> Write down any question or thoughts you may have had about your designed environment.

In Chapter 6 the introduction of national curriculum material with the emphasis on designing, making and appraising (DMA) was considered (Curriculum Corporation 1994). As a result we now see a range of technological activity occurring in classrooms and centres. However, little thought has been given to the context in which the design questions emerge. The design questions drive technological activity. A good design question facilitates quality technological learning and therefore is an important aspect of the teaching–learning process.

This chapter presents a discussion on the importance of encouraging children to ask quality technological design questions, and for teachers to use these questions as the foundation for developing their technology teaching programs. Examples are given from the book *I Can Make My Robot Dance*, published through the Curriculum Corporation (1995).

DETERMINING CHILDREN'S DESIGN QUESTIONS

As educators we want children to ask technological questions. Questions that we can use to build relevant and interesting technological teaching programs for our children. Children are capable of asking purposeful and insightful questions.

The following questions were asked by a group of four-year-old children after exploring a series of robotic toys brought in to preschool:

- Do they have switches?
- Do they have buttons?
- Do they like to eat oil?
- Do they move on wheels?
- Do they have a motor?
- Do they talk?
- What jobs do they do?
- Do they have beds?
- What time do they go to bed?
- Do they have feelings? (Fleer & Sukroo 1995, p. 168)

The level of questioning ability of these children is evident when thought is given to the question 'Do they have feelings?'. Vicky, their teacher, supported her children's learning through providing experiences, such as the Valiant Roamer, alongside of the robots they had brought into preschool, to actively explore basic programming principles. The children investigated their robots based on the questions they asked. In their investigations they wrote lists (programs) of what their robots could do, and later what they wanted them to be able to do—such as designing dancing steps. Children asked most insightful questions, such as:

> Why can't my robot remember the steps I put in it?
> How do you rewind my list (program)?

If four year olds are capable of asking sophisticated questions such as these, what is possible with eight or ten year olds? Can we teach children to ask questions of their created environment when in a school context?

QUESTIONING OUR CREATED ENVIRONMENT

If we look around the environment that we live in, it will become immediately obvious that a great deal of it is constructed by humans. We are born into particular environments and cultural practices. As children they are all new. We have many questions to ask. As adults we no longer see and question them, since our environment has become so familiar to us. It is only when we travel that we begin to think about the differences we see.

Within our own lifetime and in our own community, we find that the designs that are created actually come from within the culture and environment that exists. Most designs are only marginally different from those that exist. Radical changes in design are really only incremental.

> *Design never starts at zero, for it always starts with an already existing designed object and comes from a particular environment* (Fry 1994, p. 10).

The designed environment that we take for granted influences how we live. Road systems or town planning conceptualised over a century ago influence how we interact today. Many designs outlive the designer. Fry (1994) has suggested that:

> *Design goes before what is made and continues on after it has arrived. The implication is that the agency of design is not just the designer but also the designed* (p. 25).

If we consider an aerial view of Sydney, we can see how the town planning in central Sydney has evolved since European settlement. How people move about and live lies in contrast with that of Canberra. Canberra is a planned city. The buildings and road systems have taken account of present and future needs and population growth. Consequently, a design 'always goes on designing—unless destroyed' (Fry 1994, p. 25). Designs take on a life of their own. We work with good or bad designs. Do we question the design solutions we must live with?

Most design solutions are based upon some criteria. Fry (1994) has suggested that design solutions to particular problems or needs can enable or limit people:

> *... the design object always has an actual, or imminent, utility or sign function that* either enables or delimits a relation with it (Fry 1994, p. 25); [author's emphasis].

We need to question the thinking that supports particular design solutions.

- What is the design solution concealing?
- What limits are imposed?
- What are we now able to do that we could not do before?
- What are the implications of the design and resultant action, now and in the future?

Children question their environment. We need to build upon their curiosity—as they see things we no longer see, things we have taken for granted. Questioning the thinking that underpins design solutions is a more sophisticated form of thinking.

> JOURNAL ENTRY 10.2 *Encouraging children to ask quality design questions*
>
> How could you encourage young children to ask quality design questions?
> What could you set up in a classroom or centre which would act as a catalyst for asking design questions?
> What would you do or say?
> Record your ideas and bring them along to tutorial for discussion.

A FRAMEWORK FOR STIMULATING QUALITY DESIGN QUESTIONS

In order to question our designed environment, we need to 'make it strange'. The school context does not easily support children in asking questions of their environment. In Table 10.1 a framework for stimulating quality design questions of our environment is presented.

Table 10.1 *Questioning the environment*

Design questions	Description
What is the life of the design?	Are we living in an environment that was designed last century, such as the road system, railways etc?
How does the design influence how we interact?	If we live in Sydney (evolving town planning), as opposed to Canberra (planned from the onset), what does it mean in relation to how we live or interact with each other?
Whose interests are being served?	Are issues of equity compromised in the design solution? For example, what does it mean in terms of power, equity and inclusiveness?
Does the design address cultural sensitivity?	Does the design solution take into account differing world views?
Is the design sustainable?	Is the availability and use of scarce resources considered in the design solution?
How else could it look?	

How would this framework translate into classroom practice? Perhaps an ecodesign approach provides the foundations for unleashing children's questions. Fry (1994, p. 11) has suggested that we need to 'think ecodesign as a new and plural design paradigm'. He believes that 'the creation of educational theory, practice, methods and materials should be based on developing ecological sustainment' (Fry 1994, p. 31).

> **JOURNAL ENTRY 10.3** *Ecodesign teaching*
> How could you use the framework presented above to teach ecodesign principles?
> Map out a lesson plan on what you might do. Bring this along to class for sharing.

In the following section, some examples are presented on how very young children can achieve quality design questions. The focus is on developing ecological sustainment.

Architects and building

Vanessa, a Year 2/3 teacher, set up her classroom as an architect's studio (e.g. computer with CAD program, models, plans, drafting equipment, an architect working). Children moved around the room, taking notes and recording the questions they had. The ecodesign approach prompted the following questions and further investigation (see Table 10.2.

Table 10.2   Architects and building

Questions	Investigations
What are houses made from?	An investigation of housing around the world. Different materials and different design solutions (culture, climate, availability of resources).
How do buildings get water and electricity?	An investigation of service models; using a simple computer-assisted design program to make their own designs—designs which considered energy efficiency; devising alternatives (looking at different homes in the community to see if they were designed with or against nature, looking at the water catchment and recycling grey water).
How do cities get built?	An investigation of the impact of design (or not) and living with the artefact well beyond the life of the designer.

Restaurant

In many centres and classrooms children are involved in a range of cooking experiences. Shelly, a pre-service teacher, decided to set up a restaurant, the Jelly Fruit Restaurant, in her classroom—with the children designing, making appraising the space, food and system for eating. As part of this unit, the children also visited a local family restaurant. The questions they asked (when they visited the family restaurant) as a result of the ecodesign approach are shown below in Table 10.3.

Table 10.3   Restaurant

Considering the environmental effects	The food and the materials
What do you do with the food scraps?	Do you use free range eggs?
Do you have a scrap bucket?	Are your vegetables organically grown?
Where do you put your empty drink cans and bottles?	Do you use products that encourage driftnet fishing?
Who drives to the recycling centre for you?	What material are your chopsticks made from (in the context of the impact on rainforests)?
What kind of things do you use for cleaning?	Do you use recycled paper?

All wrapped up

Teachers can also specifically introduce the ecodesign approach. For example, Vicki, a preschool teacher, decided to deliberately investigate the 'by-products' of Christmas. She focused her attention on the excessive drain on our resources over Christmas—wrapping paper, Christmas decorations and Christmas cards. She began her unit by telling a story about three animals who wished to celebrate Christmas. The animals wrapped their gifts in different types of materials. She then drew the children's attention to ecodesign principles:

> *Would you put the wrapping paper in the bin?*
> *What else could the animals have done with the wrapping paper?*

She then asked the children to consider the following:

> *Have these materials been used before?*
> *Can we use them again?*
> *Can these materials be used in a different way?*
> *Could other materials have been used?*
> *What materials could be used again for wrapping presents?*

The children then progressed to using recycled materials to create their own decorations, cards and wrapping paper. The children also utilised the school computer to design and create their own Christmas cards on recycled paper.

Collectively these ecodesign questions can be represented in the following framework (Table 10.4).

Table 10.4 A framework for stimulating quality design questions by children

Materials	Examples
What happens to the material afterwards?	Does it go down the drain hole?
Where does the material come from?	Rainforests?
Can you recycle it?	Does it have a recycle symbol?
Do we need to use so much?	Is there a lot of packaging that is not necessary?
Does the material grow again?	Is it made of cotton (which grows) or is it a manufactured substance?
How long does it take to make the materials?	Paper: Is it made from wood which takes a lot of time to grow and to process or is it made from hemp which is easily grown and processed?
Are natural resources used?	Sunlight
Design	**Examples**
How else could it look?	
How long will it last?	Some items are designed to break after 12 months (e.g. irons).
Can you undo it and fix it, or is it a sealed unit?	Many toys are designed so that they cannot be fixed.
Are there different materials used in the product which prevent recycling?	Some items use a range of plastic and metal and make it difficult to recycle.
Have the parts which work the hardest been made of strong materials?	Axles on toy cars, tow bars on tractors, legs and arms on dolls.
System	**Examples**
Is there a recycling system?	
Do they have a recycling policy?	
Are the recycling bins easy to get to?	Where are the bins—in the school?
Are useful products made from the recycled materials?	Paper
Do they have a policy of buying recycled materials?	Tin cans or bottles
Information	**Examples**
How do we tell others about the recycling?	Designing information for presentation on computer, tape, video, diorama, notice board, maps, poster and letter.
How do we present information on the different ways of recycling?	Paper, tins, plastic, glass and compost.

SUMMARY

The basic fabric of these questions is that of ecodesign (Fry 1994). If we are able to not only listen to the questions children ask about their environment, but actively work together with them to question it from an ecological perspective, then we will develop thoughtful and caring individuals. Children are able to think about their environment, they can question it and most importantly, they have the capacity to make decisions about whether what is being presented is within the best interest of humankind. The challenge lies not with the children in our care, but with the active construction of learning environments which allow children's design questions to emerge.

REFERENCES

Curriculum Corporation of Australia, (1994) *Technology—A Curriculum Profile for Australian Schools*, Curriculum Corporation, Victoria.

Fleer, M. & Sukroo, J. (1995) *I Can Make My Robot Dance*, Curriculum Corporation, Victoria.

Fry, T. (1994) *Remakings. Ecology/design/philosophy*. Envirobooks, Sydney.

ACKNOWLEDGMENT

This chapter is a slightly modified version of a paper published in Fleer, M. (1997) 'A question of design', *Educational Computing*, 12 (1), pp. 22–5.

Chapter Eleven

Discourses in technology education

INTRODUCTION

Over 10 years ago, the Commonwealth Schools Commission released its policy on the education of girls. This document marked the coming together of two decades of research in gender education.

Over these two decades research and practice had been framed within three quite distinct paradigms of thinking: *equal opportunity, gender-inclusiveness* and *poststructuralism* (Alloway 1995). These models have directly informed policy development and classroom/centre-based practice in all fields of education. However, children are still coming to school, preschool and childcare with expectations and ways of interacting which are based on gender. Without active re-positioning of children by their teachers, young children will carry with them these values into adulthood.

> JOURNAL ENTRY 11.1 *The significance of gender*
> In the study of technology education do you think that gender is significant? Record your views.

This chapter summarises the main research literature related to gender and discusses its significance for technology education. Examples are provided to illustrate how gendered discourses influence technology learning. Broader sociological issues are discussed elsewhere.

INTERACTIONS

There is a large body of research evidence which supports the view that children have, by the time they reach school, gendered ways of behaving (Alloway 1995; Clark 1989; Davies 1989, 1994). They already know their story line—what it means to be a boy or a girl. They take to the learning situation these patterns of behaviour. As a society, these patterns of interacting and behaving are entrenched and difficult to change. In the following transcript the story line in terms of expectations of interaction become evident. A group of five six-year-old children (one boy and four girls) discuss with the interviewer what they have been doing in technology during the school term.

Researcher: 'Alison, tell me all about what you did.'
Alison: 'We had to make a map …' (she is cut off)
Alex: 'No we didn't, we writed down what we needed.'
Researcher: 'What's that called when you write down what you need?'
Alison: (puts up hand)
Alex: 'Instructions. We writed down what we needed for the puppet theatre. We thought we might be dressing up, but then we made the puppets instead.'
Alison: 'First we made …' (she is cut off)
Alex: 'First we drew the pictures then we made the plan what we wanted …'
Alison: (nodding head, agreeing, gestures to begin talking)
Alex: '… in the big area.'
Alison: 'First it was a little shop, a little home …'
Alex: 'Corner.'
Alison: (gestures to continue talking)
Alex: 'Then we took all the stuff out.'
Alison: 'Yeah. And there are blocks and stuff behind there.' (pointing and looking in direction)
Researcher: 'So you had to clear the space?'
Alison: 'We had to make the popcorn out of these little things …' (soft voice, looking in a deferential manner to Alex)

In the above transcript Alex takes control over the discussion. He expects and assumes the power to take charge over the group sharing. His expectation is 'to do the talking'. However, this expectation is not held by the girls in the group—who remain silent. Alison, who had been originally invited to share, takes a passive and supportive role. She allows Alex to take over the conversation. Any attempt to take back the discussion is done hesitantly and with very little success. The other three girls sit quietly and do not interject. Although we do find individuals (whether they be boys or girls) who interact as Alex did, the research demonstrates that it is overwhelmingly boys who behave in this way (Clarke 1989; Davies 1989).

JOURNAL ENTRY 11.2 *Observing interactions*

On practicum conduct a sociogram of the interactions occurring in the classroom or centre. An example of a sociogram is shown below.

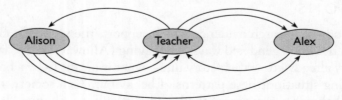

continues…

Include in your sociogram each child that initiates or supports a conversation. Use arrows for initiation and dotted lines for supportive comments. Note who does the talking and who receives most of the attention from the teacher. Take your sociograms to class for sharing.

Figure 11.1   Who takes the lead?

Alex and Alison have been constituted to operate in quite different ways. Girls and women are constituted to be supportive, nurturing and gentle, whilst boys and men are constituted to take charge, be assertive and to dominate. Their story lines are such that the girls wait and the boy, in this instance, takes charge. Had their story lines been different, their interactional patterns would also have been different. When a child has been constituted to wait and be supportive they do not assume power and take over the discourse. Being gendered as masculine or feminine enables or disallows certain behaviours, expectations and interactional patterns to emerge. Without careful repositioning by the teacher these children will interact in gendered ways. These established (and expected) patterns of interaction are generally more powerful than the educational programs which are designed by teachers to specifically address gender issue through content—for example, programs which provide books, puzzles and posters which reflect children and people in a variety of roles, clothing or nurturing/active situations. Even stories which specifically deal with issues of gender stereotyping, such as *Princess Smarty Pants* by Babette Cole, have been found to be actively resisted by children (Davies 1989). Without tackling basic interactional patterns girls' and boys' educational opportunities will be reduced and understandings gained from these educational experiences may only be learnt vicariously (Clarke 1989). For example, girls may have limited opportunities in which to manipulate construction materials or tools; or believe that their future lies in 'getting married' rather than actively pursuing a career. Boys may have limited opportunities in which to participate in nurturing activities, only learning vicariously to be gentle or caring. The potential of each individual child is not realised when these discriminatory patterns of interaction are maintained in schools and early childhood settings.

> JOURNAL ENTRY 11.3 *The teachers' role in changing interaction patterns*
> What can teachers do to increase the opportunities for both girls and boys in technology education so that individual potential is realised? Record your views.

POSITIONING CHILDREN IN WAYS WHICH ALLOW INDIVIDUAL POTENTIAL TO BE REALISED

In direct contrast to Alex's and Alison's patterns of behaviour is the discussion that is detailed in the following transcript. The interactional patterns between the boys and girls are quite different. This transcript demonstrates the way in which the teacher repositions her children to allow girls to assume a dominant role. Such examples, if sufficiently plentiful, and if boys are not unduly disempowered in the process, may well serve to contribute to balance across the scope of the program. In this example, the girls are not silenced, their voices are heard and they have expectations of being heard. What they say is constantly reinforced as being valued and intellectually important. As a result, the interactional patterns are balanced, and the girls make regular and confident contributions on a topic, robotics, that is generally perceived to be of greater interest to boys.

Context: The transcript below was typical of the interactions that were occurring at one particular preschool. The teacher, Vicky, asked individuals by name to contribute. They were required to think through their response and were given sufficient time to respond without interruption by other children. The teacher's comments indicate her valuing of the boys' and girls' ideas and suggestions.

Teacher: 'We need to do a little bit of talking and sharing about our robots. Let's start with the duck robot. Who was working on the duck robot?'

Children: 'Me!' (chorus)

Teacher: 'What can you tell us about the duck robot, and what did you find out? Hands up.'

Libby: 'It went backwards and when I pressed a button its eyes lighted up.'

Teacher: 'So you made the robot do two things. You made it go backwards and made its eyes light up. So how did you do that?'

Libby: 'I pressed some buttons.' (several boys lean into middle of circle)
Teacher: 'Which buttons did you press to make it go backwards? Can you say the words? Luke and Mark, can you sit back, because now Libby's having her turn, and then we'll show the others. Luke, own space please. So Libby what button made it go backwards?'

Here Libby is positioned in a way which re-affirms that she has a right to share her views. She is given physical space to do this and allowed time to express her ideas. Luke and Mark are positioned to wait their turn and to give Libby space.

Libby: (kneeling up, points to a button)
Teacher: 'So, you pressed this button. And what does this button have in the middle of it?'
Libby: 'A triangle.'
Teacher: 'A triangle, that's right. And when you pressed the triangle, what else did you have to do to make it go?'
Libby: 'Just pressed it and it went backwards.'
Teacher: 'So it went by itself. When you pressed the triangle, what else did you have to do to make it go?'
Libby: 'Just pressed backwards'
Teacher: 'So you just pressed the backwards button and it went backwards by itself. Can anyone else who used the duck robot tell me anything about it?'
Libby: 'You press this button to make its eyes light up.'
Teacher: 'This one here?' (pointing to it)
Libby: 'The circle one—the red and green one.'

Figure 11.2   We can make our Valiant Roamer push blocks!

In the above transcript it is evident that the girls had been repositioned to contribute without interruption. The boys had been repositioned to wait, not interrupt and provide space for other children to contribute. Any expectation by the boys of taking charge of the conversation was not reinforced. They were silenced when they attempted to take control of the discourse. In this preschool the children's interactional patterns were not those normally seen. Each child expected to be heard, to have access to resources and to be given uninterrupted time to contribute—quite a different situation to that discussed earlier with Alison and Alex.

Teachers have a significant role to play in reviewing what it means to be a boy or a girl. Boys and girls are more alike than they are different. Yet the defined roles for each in terms of masculinity and femininity are quite marked. Helping children to understand themselves as individuals disassociated from their gender will mean that they see themselves and others as people—one small step towards combating the binary or dualistic thinking so prevalent in contemporary society. For example, a two-year-old child was heard to say to another five-year-old child who had counted and labelled all the children in terms of their gender: 'Me not a boy, me a person!'. He later corrected his teacher at group time: 'That's a snow person!' (rather than a snowman).

The reinforcement from birth that someone is a 'good' boy/girl or 'naughty' girl/boy each time that person does something that is approved or disapproved of by adults builds up an image of what it means to be a boy or a girl. Similarly, the silences by adults act as reinforcements also. Each time a boy takes charge over a situation, overriding other children, particularly girls, without adult intervention, two messages are given. First, boys are given permission to dominate and second, girls must accept this. Because this is still the norm in many sectors of contemporary society, many adults do not intervene. Adults have also been socialised into expecting this type of behaviour. Hence inaction by adults is just as powerful as action in giving messages to children. Children are, after all, only trying to make sense of the social patterns they observe. Children learn social patterns through adult reactions and interpretations of each interaction. They take on the social norms people have adopted as a society—that is, they actively *take up as their own* the discourses through which they are shaped. Without contesting gendered interactional patterns, they continue.

MacNaughton (1997a) argues that this focus in interpreting children's interactions shifts the blame from the girls or the boys to the discourses through which they have learnt to understand what it means to be a boy or a girl (p. 65).

Davies (1991) has suggested that children need to be empowered to recognise the story lines that they have adopted or are being shaped into, and resist these. She argued that children should have *agency*; that is, 'the freedom to recognise multiple readings such that no discursive practice, or positioning within it by powerful others, can capture or control one's identity' (p. 51). Children need to be given access to many story lines. They should not have to take on story lines that render them powerless, but rather be encouraged to contest them. For example, the labelling that children often direct towards each other: 'technology is not for girls' or 'you are a girl' is demeaning of girls and women. Girls need to be provided with discourse that allows them to contest these negative or disempowering story lines, such as: 'Yes, girls are clever, yes, they use technology and act technologically, so it is nice to be called a girl'.

Davies (1994) also suggested that educators should ask themselves the following questions in assisting in the observation and analysis of their teaching practice:

- How are the children *positioning* each other in that context?
- Where does the *authority* lie?
- How is *experience* made relevant?
- What *binary* or *dualistic* thinking is evident in the discursive practices (e.g use of terms such as man/women, boy/girl)?
- What forms of femininity and masculinity are allowed/disallowed?
- What *story lines* are being made relevant?
- Whose *interests are being served*? (Adapted from Davies 1994, p. 45).

These issues are considered further by MacNaughton and are worthy of analysis. In her study of preschool children she found:

> *The girls regularly exercised their power through various avoidance or non-involvement strategies. The effect was that they constructed block play as a 'no-go' area when boys were present and created alternative spaces in which particular ways of being female were dominant ... They (girls) were not interested in playing in an area in which 'macho' ways of being male were the dominant ways of being* (MacNaughton 1997, p. 64).

MacNaughton (1997b) found that the boys exercised power physically in the block area and through the story lines they used. They created and maintained block play in ways which demonstrated the dominance of males.

The implications for educators who work with children, particularly in settings where children are encouraged to work and play with construction materials are enormous. MacNaughton (1997b) argues that educators need to:

- *find ways of challenging the patriarchal power order in which boys are always powerful and girls powerless; and*
- *create gender discourses in which power is more evenly distributed.*

How do these ideas relate to technology education? In the following section an example is shown of two technology education programs which introduce technological tools and materials in order to increase girls' technological capabilities. (Further discussion on this topic can be found in Fleer & Hardy (1996) *Science for Young Children*.)

Scenario One: The tinkering table

Context: Twenty-five preschool children were provided with a tinkering table. On the table were screwdrivers, pliers, hammers and saws. There was also an array of technological artefacts (e.g. telephone, keyboard, radio) and containers to put dismantled pieces in. The teacher argued that it was important for the girls to be given opportunities during free time to use tools.

An analysis of what was happening:
During free choice times there was always at least two boys who stayed and pulled apart the technological artefacts. Each boy remained for at least 15 minutes. Usually one girl would

join the group briefly, mostly observing what was occurring and then move on. An example of the conversations held by two girls who approached the tinkering table follows:

Sam: 'I'm going to undo this (keyboard).'
Peter: 'I might undo that one (telephone).'
Sarah and Freya approach the tinkering table.
Sarah: 'They have done a lot! How did you get the telephone out?'
Peter: 'I didn't, it was already out, and now I am breaking this (backing) off. You're allowed to break it.'
Sarah: 'So we are not *breaking* it are we?' (said with a strong emphasis)
James: 'Well, that's a way of breaking it! Yep, you're doing it undone.'
Sarah: 'Have you done a painting?' (addressing Freya who is observing) 'Come on, you can come with me and do a painting.'

Both Sarah and Freya move away to the painting area. The remaining children work on in silence.

(Modified from Fleer 1990, pp. 360–1)

It can be speculated that the girls saw the tinkering activity as destructive, since it was viewed as the breaking apart of an item. Here the girls' comments and subsequent lack of participation would give support to this idea. It is interesting to note that the girls predominantly watched the boys tinker and then moved off to other activities within the preschool.

Scenario Two: A study of clocks
Context: A Year 2/3 classroom. The children are grouped into self-selected single-sex groups of two or three. The teacher had planned a 10-week program focusing on how clocks work. The children were encouraged to dismantle the clocks and ask questions about the technology.

An analysis of what was happening:
The girls' conversation concentrates upon fixing or constructing. The following transcript illustrates the constructive nature of the technology.

Andrea: 'Well, we're going to make a robot now.'
Researcher: 'You're going to make a robot now are you?'
Andrea: 'Yes, umm.'
Helen: 'The earrings.' (takes another piece from the clock)
Researcher: 'How are you going to make it?'
Andrea: 'I don't know. If you put this thing on here, and then she got this bit and then she was trying to bend this bit …'
Helen: 'No, she's not in our group.' (looking at a child who hovers nearby)
Andrea: 'We need to put her ears on.'
Helen: 'And her earrings.'
Samantha: 'Ears.'
Helen: 'Earrings?'

Andrea: 'Yeah, she can have earrings.'
Samantha: 'She'll be a rattly tummy.'
Andrea: 'The arms.'
Samantha: 'Here are the arms.'
Helen: 'They look like big dangly earrings.'
Andrea: 'The arms.'
Samantha: 'Umm, what were her legs?'
Helen: 'I know!'
Samantha: 'And this goes …'

(Modified from Fleer 1990, pp. 361–2)

Here the girls considered the task in terms of the construction of a robot. They examined the clock not for the identification of clock parts or for an understanding of the clock mechanisms, but for the selection of pieces to use in the construction of a robot.

In contrast, the boys were preoccupied with pulling the clock apart and bringing to the task either previous destructive experiences or focusing on how to dismantle each component of the clock without regard for reason or purpose, as is evident in the following transcript.

Chris: 'We have finished our clock, we are taking apart this radio, we are trying to work out how to take off these screws, so then the teacher comes and unscrews one and then we start, then we take it all out.'
Researcher: 'What's this here?'
Chris: 'We don't know, but I just unscrewed that.'
Researcher: 'What do you think it's for?'
Chris: 'Oh, that's most of the radio. Look at this.' (moves numbers) 'We are having quite a lot of fun. My little brother thinks these are bombs.' (pointing to white cylinder objects inside radio section). 'We are going to take these apart and see what we can do with them. I had a radio like this at home and my friend came up, we both like taking apart things, and we made bombs and those can explode.'
Researcher: 'They can explode?'
Chris: 'Yeah, we put them on to wires, and put some of that on, and all these little things, and they did explode, we lighted them outside, at night and they really made a big explosion.'

(Modified from Fleer 1990, pp. 362–3)

The boys brought to the learning situation their prior experience in dismantling machines. Their frame of reference in attending to the task related to the identification of components in other technological items. Here they used their prior ideas (bombs) and previous experience (causing components to explode) to help them understand the new experience (the inside of a clock).

A sharp contrast in language was evident between the boys' groups and the girls' groups. Rarely did the girls engage in destructive language and similarly infrequent constructive comments were heard within the boys' groups. The boys were bringing to the task prior experience which embellished their language further, and perpetuated their home-based experiences.

It is interesting to note the contrast between gender groups, in how children framed their task, namely a constructive framework and a destructive framework for thinking about the task at hand.

When the literature on girls' preferences to learning is considered, it would seem that a disparity arises, as the destructive element seen in the tinkering activity by the girls in the preschool and the Year 2/3 classroom is contrary to girls' preferences for creating/helping or relating to people. In the classroom, the activity was re-framed; for example, making a robot, to help girls engage in, and make sense of the destructive task. Indeed, in the preschool the destructive language used by the boys and the nature of the task deterred girls from actively participating when given the choice.

By placing the material (old machines, clocks) and the equipment to pull the components apart (screwdrivers and pliers) on tables without adult guidance or interaction ensured it became a solitary activity. As a result, it reinforced the notion that technology is about 'things, not people' (Kelly 1987). It also decontextualised the activity as it provided no purpose in the task other than destruction.

This was particularly evident in the classroom. The children attempted to contextualise the task by associating it with previous learning: in the case of the boys, with bombs; in the case of the girls, the construction of a person, namely a robot. Even the notion of seeing what was inside the particular machines was of no consequence, since the complex nature of the item (cassette player, telephone, clock) meant that the particular components did not relate to their present level of scientific and technological understanding, resulting in a limited comprehension of what they saw and consequently little chance that they would ask scientific or technological questions which they could investigate further. As a result, the majority of the children's classroom discourse was not related directly to the task at hand (as determined by the teacher).

Over the course of the 10 lessons in the Year 2/3 classroom, the children's knowledge of what was inside the clocks developed. However, the relationship of components to each other or a firm understanding of the operation of a clock or even the naming of components was limited. Adult intervention was predominantly managerial and hence procedural. No framing of the task into a social context occurred.

The technological context: a constructive or destructive focus

Each teacher's rationale for including tinkering in their program was to first allow the children to see what was inside the machines, second, to provide them with opportunities (particularly girls) to use screwdrivers and work with mechanical things, and third, to give the children the opportunity to ask scientific questions about elements of what they were tinkering with. However, through the representation of this activity, it tended to favour the boys' preferences of learning above the girls'. The decontextualised nature of the experience in combination with the destructive focus, did little to attract or hold the interest of the girls.

The danger here is that with a new area being introduced into the curriculum, namely technology, it too may be presented in a socially decontextualised manner. As was evident from the transcripts, when children were presented with technologically-based tasks they either chose not to become involved, or created their own contexts. The creation of their own contexts was based on their prior experiences, which were gender-focused. The contexts created by the children themselves did little to assist them to develop an understanding of the technological materials they were tinkering with. The context created by the children reinforced the gender divisions, and when the girls were given the choice not to participate, the technology activity transformed into a male activity.

It has been argued by MacDonald (cited in Kelly 1987) that gender is often re-contextualised within school, so that what is deemed as appropriate behaviour for boys and girls is converted into what is appropriate behaviour for boys and girls in particular curriculum areas. Kelly (1987) goes on to say that the gender-differentiated ideologies and behaviours:

> ... *do not necessarily have any direct relevance to school subjects such as French or physics* (or technology). *But new differentiations can be linked to existing ones, so that some school subjects come to be seen as masculine and others as feminine ... Thus once a subject has acquired a masculine image, participation in it is seen to enhance a boy's masculinity and diminish a girl's femininity* (p. 68).

With the introduction of technology into schools and preschools, careful thought needs to be given to planning, otherwise the male image may well be fostered and girls will be faced yet again with curricula that favours the boys' interests and learning styles. The teaching of technology is still a relatively new area in the curriculum (although it is older than science in reality), consequently there is still time to contextualise it socially. This contextualisation will not only favour girls' learning styles and prevent this area from taking on a masculine image but ensure that children, when engaged in technological tasks, are engaged in an experience that develops their technological understandings.

When technology is introduced within the constructivist notions of child-directed experimentation and a non-directive role for the teacher, children often react in ways that are contrary to the teacher's educational purpose. In the transcripts shown above, the children's behaviours in relation to the technological artefacts they were exploring reinforced rather than modified gendered interaction. When a choice of participation was offered, girls largely avoided the activity and acted as passive observers. Boys, while they did participate, constructed contexts for participation that reinforced aggressive and destructive orientation to the task. When participation was made compulsory, both groups responded by constructing their own context for the task.

JOURNAL ENTRY 11.4 *Developing girls' technological capabilities*

The introduction of the tinkering table in the preschool and the sequence of technological learning over 10 weeks with clocks were designed to enhance girls' technological capabilities. Yet this goal was not achieved. How could the teachers have met this goal? Record your ideas.

DISCOURSE AND TECHNOLOGY TEACHING

In the past, educators have adopted simplistic approaches to gender equity programs in preschools and schools. For example, MacNaughton (1997a) suggests that educators cannot bring about change by placing 'girl-friendly' materials such as decorative items in the block corner or by placing 'boy-friendly' items such as tools in the home corner. This method does not get at the fundamental gender-based interactions that will occur in either play space. In MacNaughton's (1990b) study, those teachers who used this approach (termed 'feminisation') found that the boys moved into the space before the girls and simply moved all the feminine artefacts to the side and continued to play as they always did. This perspective is also held by Davies and Banks (1992), who reported that:

> ... *equity programmes which simply introduce the idea of equity, and which rely on role models and access to non-sexist curricula, will not be enough to disrupt these strongly held theories of gender and patterns of desire* (p.23).

MacNaughton (1997a) suggests that educators also have attempted to introduce change though separating boys and girls during free play. For example, many schools have introduced days in which only girls are allowed on the school oval. Similarly, many preschools have adopted girl-only block play days or times. MacNaughton concluded that when girl-only times ceased, the children reverted to playing in their 'old' manner. This approach did little to challenge current interaction patterns or beliefs about gender and power held by the children. This position was reaffirmed by observations made by MacNaughton (1997a) in the block corner of boys actively seeking to disrupt the girls' play during the girl-only times.

The assumption implicit within the provision of girl-only times or the introduction of girl-friendly materials is that the problem lies with the curriculum materials and the girls. The discourses that are used to shape play or the story-lines that are taken up by both boys and girls are not challenged.

Another common approach to increasing girls' involvement in play which is usually the domain of boys is through active policing by adults who are supervising the play environment. MacNaughton (1997a) found this approach highly successful in allowing access and maintaining play for its natural duration. However, she also found that once the adult was called away (for example, to answer the phone or attend to an injured child) the girls' play was disrupted or taken over by the boys. Gender relations were not only policed by the adults, they were the adults' responsibility, leaving limited opportunity for the children to actively contest the story lines. The assumption implicit within this approach is that boys are defined as the problem.

Another approach, not commonly seen in schools or centres is known as 'fusion'. This involves combining an area of play predominantly utilised by girls, such as the home corner, with a play space occupied by boys, such as the block area. MacNaughton (1997a) reported that when this approach was used the following story lines were adopted and seen as normal:

- solving problems via physical aggression;
- killing and 'getting' the bad guys;

- capturing space;
- being loud; and
- touching each other physically in 'rough and tumble' play (p. 60).

MacNaughton (1997a) found that the girls' story line involved:

- being 'mums';
- cooking and cleaning;
- shopping;
- looking after baby; and
- having parties.

She indicated that when the boys' play and the girls' play intersected, the girls would either continue to play around the boys or move out of the space altogether. The assumption underpinning this approach is that the materials and the children themselves are the problem.

The simplistic approaches adopted by practitioners in the past are based on a limited understanding of the discursive practices that operate within society. Educators need to work together with children to:

- understand the concept of discourse;
- identify the dominant discourses;
- recognise the discourses of resistance;
- analyse storylines; and
- become tuned into the positioning that takes place during interactions.

In summary, Davies and Banks (1992, p. 23) suggest:

> *To deal with this contradiction we suggest they need to understand how they (the children) are interpellated, they need to understand how they have taken up various discourses as their own, and how desire is implicated in their preferred story-lines. They need as well to grasp the contradictions that inevitably exist and which are not allowable in the humanist framework in terms of which most of them think. They need to understand how discourses of resistance work, if they are to begin to engage in a radical personal change which undoes fundamental elements of the male/female dualism.*

Educators introducing technology to children must focus on discourses that not only surround technology but those that are gender-based. Otherwise teachers will ask:

> 'Even though I give the girls every opportunity in technology education, they still do not participate. Why don't the girls do technology?'

Teachers should be asking:

> 'What gendered interactions are occurring and who is exercising power? How can I change the discourses in technology education lessons to ensure that power is more equally distributed?'

It is important for teachers to examine the technological lessons they are providing and critically evaluate how much of children's experiences are fundamentally about gender relations. As MacNaughton has argued:

> *We need to shift from a focus on how many boys and girls are in block play to questions of how they are playing and who benefits from the play that takes place* (MacNaughton 1997a, p. 66).

SUMMARY

Teachers have a significant and timely role to play in contesting gendered interactional patterns and developing programs in technology which build upon children's experiences in appropriate ways. Whilst it is difficult to completely change what society is continually reinforcing, it is possible through repositioning young children to provide more balance and promote equitable interactional patterns—albeit for only a limited time each day. This provides another story line for those children to consider and it allows the girls to be 'talked into existence' for at least the time they are at school.

REFERENCES

Alloway, N. (1995) *Foundation Stones. The Construction of Gender in Early Childhood*, Curriculum Corporation, Melbourne.

Clark, M. (1989) *The Great Divide. The Construction of Gender in the Primary School*, Curriculum Development Centre, Canberra.

Davies, B. (1989) *Frogs and Snails and Feminist Tales. Preschool Children and Gender*, Allen & Unwin, Sydney.

Davies, B. (1991) 'The concept of agency: a feminist poststructuralist analysis', *Social Analysis* 30, p. 51.

Davies, B. (1994) *Poststructuralist Theory and Classroom Practice*, Deakin University Press, Deakin University, Geelong.

Davies, B. & Banks, C. (1992) 'The gender trap: a feminist poststructuralist analysis of primary school children's talk about gender', *Journal of Curriculum Studies*, vol. 24, no. 1, pp. 1–25.

Fleer, M. (1990) 'Gender issues in early childhood science and technology education in Australia', *International Journal of Science Education*, vol. 12 (4), pp. 355–67.

Harding, S. (1986) *The Science Question in Feminism*, Open University Press, Milton Keynes.

Kelly, A. (1987) 'The construction of masculine science', in *Science for Girls?*, A. Kelly (ed.), Open University Press, Milton Keynes.

MacNaughton, G. (1997a) 'Who's got the power? Rethinking equity strategies in early childhood', *International Journal of Early Years Education*, vol. 5 (1), pp. 57–66.

MacNaughton, G. (1997b) 'Feminist praxis and the gaze in the early childhood curriculum', *Gender and Education*, vol. 9 (3), pp. 317–26.

ACKNOWLEDGMENTS

Sections from the following publications have been incorporated into this chapter:

Fleer, M. (1998) '"Me not a boy, me a person": Deconstructing gendered interactional patterns in early childhood', *Australian Journal of Early Childhood*, vol. 23 (1), pp. 22–8.

Dockett, S. & Fleer, M. (1999) 'Pedagogy and play in early childhood education: Bending the rules', Harcourt Brace, Sydney.

Fleer, M. & Hardy, T. (1996) *Science for Young Children: Developing a Personal Approach to Teaching*, Prentice Hall, Sydney.

Special thanks to Vicky Bresnan and her children from Duffy Preschool for being involved in the research project.

Chapter Twelve

Cooperative technological learning

INTRODUCTION

In Chapter 10 you found out that we are all designers, using design practices every day. When we talk to children who have been doing technology in primary school they often talk about 'making' things and deciding on their designs. The two aspects, 'making' in a DMA approach, and cooperative learning in groups, are the foci of the present chapter.

Researcher: 'What do you actually learn in technology at school?'

Kirstyn: 'We learn how to work with people and how to design things. I thought it was really fun and I've learnt how to build things by going into all the details.'

Emma: 'We actually get to make things in technology and learn. It was good to learn about the cranks and gears. We learnt to solve problems—instead of wood use cardboard; use a couple of other bits of wood to make it stable.'

Mary: 'I think the toy would be my favourite or the bird feeder because the bird feeder was put to good use. I've still got my bridge which I like a lot, making it was good fun. I like having the chance to use the tools and making things and learning about the different parts.'

Liz: 'We had a look through Mary's book of bird feeders and first of all we decided on a house but then that would have been too hard to make. So we thought to have like a square, put some containers on the top and hang it from wire because I've got one at home a bit like that but it's got a thatched roof and things, so we decided that it would be a square and all the wire would come up into the middle and into the top. So we all made our own design of what we thought it would look like and then we decided which one it would be and then each person did one from a different angle. Because then it might be easier to follow on the plan if we could see it from a couple of angles.'

For these children, the designing, making and appraising of bird feeders, bridges and toys throughout the year were their first school experiences of technology. Their teacher chose to organise the children so that they worked in small groups formed in various ways. For the bird feeder activity the children worked in friendship groups, whereas their names were drawn out of a hat to form the groups to make their bridges. Does the group make a difference to the children's approach to learning?

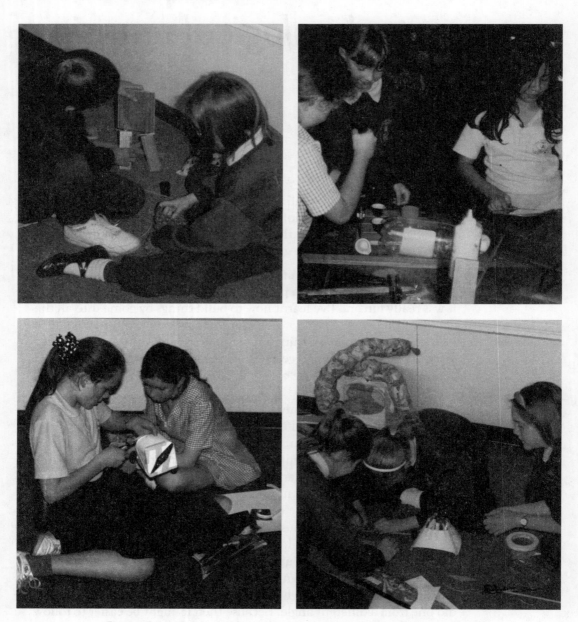

Figure 12.1   Group work is an important skill for working technologically

EFFECT OF THE GROUP

Let us further consider the bridge-building class that the children were in. Some children saw the bridge task as a social process of interaction within the group, as well as designing the bridge to be tested. For a number of children, who they worked with made a big difference to how they experienced the task and their understanding of it. In several groups different aspects of the task were carried out by particular group members. Below children talk about the group interactions and some of the difficulties that occurred due to the social element.

Researcher: 'Did that make a difference, the group you were to work with?'

Jeremy: 'I think so. I only like working with my friends as you enjoy it more. Each member did their design and we picked out an aspect from everyone. You just did your part and that was all.'

Karen: 'I worked with Jim, and Moira was sick all the time and Jim was just clowning around, so I basically did it. It was hard to make it. I found it a little bit easier to use a different design, so I just drew one up and sort of copied off that.'

Sam: 'I didn't really enjoy the bridge activity that well because who we got to work with was drawn out of the hat. I didn't really enjoy it very much because it was two girls and me and they sort of did most of the work. The bird feeder was fun, I wish we had have got it finished though. Marble mazes are fun. I like making things. It's fun because Miss Jobling is the first teacher that we've had that does things that you have to build them with nails and all that.'

As you can see by Jeremy's comments above, he believes that the group does make a difference. He had to contribute to the bridge design and was actively involved in the whole process, which is characteristic of a deep approach (which leads to understanding). In contrast he adopted a surface approach when he was with his friends making the bird feeder, and when he worked individually on his toy. Jeremy chose to work with cardboard for the toy and 'likes to make basic things that aren't fiddly'. His focus was restricted to the materials and the making aspect, which indicates a surface approach to the task.

Both Karen and Sam did not like the bridge task because of the group, with the reason being that some group members did not have the same commitment to the task as they did. When children are committed to collaborate, they try to make sense of each other's interpretations of the situation and engage in mutually supportive activity. Peer group interaction encourages an exchange of viewpoint and verbal elaboration but the quality of the verbal exchange depends on the willingness of the group members to contribute.

DEVELOPING COOPERATIVE SKILLS

Their teacher, Wendy Jobling (who you read about in Chapter 5), intended that the children should develop cooperative skills by working in groups, but this takes time and effort.

Wendy: 'Sometimes the teacher needs to help sort out any of those social problems that arise because of the random selection of group members, like arguments, and get the children to negotiate, compromise and learn all those skills, rather than the teacher imposing the solution to a particular problem.'

Researcher: 'That's another point, the children all cooperate don't they?'

Wendy: 'But that's taken quite a bit of development.'

Small group problem-solving provides opportunities for learning that arise from collaborative dialogue as well as from the resolution of conflicting points of view. In addition to contributing their ideas and skills towards achieving a successful product which works, the children also learn personal skills which can be transferred outside the classroom. In order to successfully complete the group's product the children have to cooperate with one another.

When we visit schools we see most primary school teachers implementing technology by organising the children into small cooperative groups. An important social skill that is fostered from early childhood onwards by teachers is the ability to work effectively with others. Cooperative learning extends on the informal cooperative work that is common in early childhood settings.

A cooperative learning group is 'a group in which students work together to accomplish shared goals. Students perceive they can reach their learning goal if, and only if, the other group members also reach their goals' (Johnson & Johnson 1996, pp. 1–16). In this chapter you will investigate cooperative learning strategies for small groups by participating in cooperative learning activities yourself.

Think back to popular children's books that you enjoyed which may be useful for setting up design challenges. *Gulliver's Travels* adventure story is the example which prompted the following challenge. Gulliver set sail from Bristol (an English port) but encountered a violent storm to the north-west of Van Dieman's Land which caused him to be washed up on the beach. Exhausted from being tossed around by the waves he lay down and slept for nine hours. On wakening at daylight he recalls:

> *I attempted to rise, but was not able to stir: For as I happened to lie on my Back, I found my Arms and Legs were strongly fastened on each Side to the Ground; and my Hair, which was long and thick, tied down in the same Manner, I likewise felt several slender Ligatures across my Body, from my Armpits to my Thighs.* (Asimov 1980, pp. 9–10).

He had been pegged down by the little people of Lilliput.

Figure 12.2   Gulliver with his hair pegged down

Making straw towers in Lilliput Land

Imagine you are an engineer in Lilliput (home of the little people) and that you have been asked to submit a design for a tower to support an apple at least 15 cm above the ground. The tower is to be made of straws joined together by pins or pipe cleaners.

The challenge is to build the tower using as few straws as possible. Each straw costs Lill. $4, and each time you cut a straw it costs Lill. $1. You may use small amounts of blue-tack to stop your straws slipping. A piece of blue-tack the size of a pea costs Lill. $2.

The tower must support the apple for at least 30 seconds (time flies in Lillput!).

(Tytler 1990; Klindworth & Jane 1993)

Figure 12.3   Design brief

For safety, preschool children could use the pipe cleaners placed in the straws as a method of joining them together.

JOURNAL ENTRY 12.1   *Making straw towers in Lilliput Land*

In a group of three, assign roles (recorder, reporter and gofer) to each member of the group. Your task is to work as a cooperative group, with the *gofer* collecting the materials and equipment, the *recorder* writing down the materials used and their cost, and the *reporter* reporting back to the whole tutorial about the group experience of the task. As a group, build one tower and then answer the following questions.

continues...

a) Did you explore the materials first (e.g. try out different joining methods before starting to make the tower)?
   This stage in a technological process is important as it forms the investigating phase. You should allow children sufficient time to explore the properties of the materials they are working with.
b) Did your group draw a rough sketch or design prior to making the tower?
c) How did you decide which member's idea was to be used for the tower?
d) How did you evaluate your model?
e) Did the cost of the materials significantly influence your design, or did you not worry about how expensive your model might be?
f) Do you think the costing of the tower contributed to the production of a quality product?

Processing group functioning
Determine the quality of your group cooperation by talking about how well the group worked together. Each member is to fill out the following partnership processing form (Johnson & Johnson 1996, pp. 5–10).

Partnership processing form
1. My actions that helped my partners learn:
a)
b)
c)

2. Actions I could add or improve on to be an even better group member next time:
a)
b)
c)

Have your reporter share your group's experiences of this activity with the tutorial group.
   Reflect on the making of the straw tower using cooperative groups, and the potential use of this activity with children. Write down any advantages and disadvantages of the activity, along with any interesting ideas that come to mind.

Cooperative skills have to be learnt and developed for groups to function effectively. The partnership processing form you have just used can assist children to reflect on their contribution to the small group.

This same challenge (but using a different cultural context) was given to teachers from several Asian countries during a course they were undertaking. They found that it wasn't as easy as they thought it would be to make a stable structure. Only one group sketched their design before making it. Most groups used trial and error, and changed their design as they tested out the strength of the tower. Some had trouble making a stable base. All groups discovered that a triangular shape was strong and that struts provided extra strength. Some forgot to measure the height until the end, and then they had to modify the height to conform to the design brief specifications. Most importantly they were all successful and they enjoyed the technological experience. The photographs in Figure 12.4 show that the teachers were proud of their solutions.

Figure 12.4 Teachers in Malaysia making straw towers

Above we have outlined a model for setting children challenges which are contextualised (i.e. finding a good story, reading it to the children and then talking about the content in such a way that a design brief is created). You may know someone who has a collection of models of authentic English cottages, castles and inns which were made at Lilliput Lane in Penrith, England. At the Visitors' Centre for Lilliput Lane people can see the stages of a technological process which produces the intricate models. Bringing some of these models into the classroom can also help set the scene for the straw challenge.

BASIC ELEMENTS OF COOPERATIVE GROUPS

When children are placed in groups to work in their own way they may display poor learning outcomes, with aggressive children dominating and less confident children being merely spectators. For cooperative learning to occur groups have to be set up in a certain way. What is cooperative learning?

> *Cooperative learning is a learning process in which students work in small teams of mixed membership on activities requiring the exercise of collaborative social skills and task demanding the combined efforts of students to achieve both individual and group learning goals* (Waterworth & Shepherdson 1993, p. 44).

The essential elements needed to foster a cooperative learning environment are:

Positive interdependence or team rewards
Children must perceive that they need each other in order to complete the group's task—'sink' or 'swim' together. Several aspects of the task should be interdependent, such as the learning goals, joint rewards, shared resources and assigned roles.

Face-to-face interaction
Children promote each other's productivity by helping, sharing and encouraging efforts to produce. Members explain and discuss what they know to the group. Teacher structures group so that members sit knee-to-knee and talk through each aspect of the tasks they are working to complete.

Interpersonal and small group skills
Social skills are important in cooperative learning and children must be taught to use them in their groups. Collaborative skills include decision-making, trust-building, communication and conflict-management skills.

Group processing
Children should be given opportunities and procedures to analyse how their social skills can work to assist all group members to have effective working relationships which can be maintained.

Individual accountability
To be able to contribute, support and assist one another, every child should master the assigned material. Children should be challenged to give their best contribution by improving on their past performances (Davidson 1990; Johnson & Johnson 1987, 1997; Slavin 1995; Van der Kley 1991).

Having children work in groups means that it is difficult to assess each individual's contribution to the group. It is therefore important that the teacher monitors individual achievement as well as group achievement.

SETTING UP DIFFERENT MODELS OF COOPERATIVE LEARNING

In the straw challenge you participated in one model of cooperative learning which involved informal cooperative learning groups. Below we outline how you might set up other models of cooperative learning, such as Jigsaws (Slavin 1995), Student Teams Achievement Division (STAD), Timed Pair Share and Round Robin (Kagan 1992).

Jigsaws
1. To set up the Jigsaw 1 model, present children with a topic which is clearly defined into sections. Within each 'home' group, each member has the initial responsibility for mastering one section. One member from each group who has been assigned the same

section to master, gets together and 'specialist groups' are formed. In the specialist group members help one another learn a section, and they become experts on that part of a topic. The children then return to their 'home' groups and teach the material to other members of the group.

2. The Jigsaw 2 model is similar to Jigsaw 1 except each child is given all the sections of the topic.

Student Teams Achievement Division (STAD)

This model involves setting five major components: class presentations, teams, quizzes, individual improvement scores and team recognition.

1. The teacher introduces the material in a *class presentation*.
2. *Teams* of four or five children form heterogenous groups to prepare the members to do well in the quizzes. Children discuss problems together and compare answers for the good of the team.
3. The children take individual *quizzes* where every child individually is responsible for knowing the material.
4. Each child is assigned a 'base' score derived from previous performance on quizzes and gains points for the team when quiz scores exceed the base score. The team gains points only when the children show *individual improvement scores*.
5. *Team recognition* is gained through rewards such as certificates if their average scores exceed a certain criterion.

(For further details refer to Slavin 1995; Jacobs 1995)

Timed Pair Share

Children share with a partner for a predetermined amount of time and then the partner shares with them for the same amount of time.

Round Robin

Children in teams take turns talking.

Informal cooperative learning groups for technology

Assign the children to the groups with a design brief which sets the parameters of the *task*. Establish *positive goal interdependence* by asking for one design drawing from the group that all group members contribute to and can explain. The *criterion for success* is the completion of a product that works. Ensure *individual accountability* by observing each group, making sure that any one member can explain any part of the product at any time (sharing with the class). Inform the children that the expected social skills to be used by all children are:

- encouraging each other's participation;
- contributing ideas; and
- summarising.

While children work in their groups the *monitor* systematically observes each group and intervenes to help with interpersonal and small group skills. Group members then *process* how well they worked together by identifying actions each member engaged in that helped the group succeed and suggesting one thing that could be added to improve their group next time. Their product is *evaluated*, which may involve orally presenting their product to the class.

> JOURNAL ENTRY 12.2 *Cooperative learning outcomes*
>
> List other ways of cooperative learning you may have seen operating whilst you have been on practicum or during teaching experience. What other ideas do you have concerning cooperative learning? Write in your journal what the outcomes might be for children.

Cooperative learning may vary for different age groups. In the 0–3 age group, cooperative learning involves the staff helping the whole group interact with each other; in preschool, cooperative learning can be seen during free play in the home corner, outdoor area and construction area—teachers could structure this at specific times to include a 'group leader and a recorder'. In the first years of school, where children are expected to work together more formally the suggestions made above are appropriate.

OBSERVING CHILDREN IN GROUPS

Participation in technological activities can foster positive attitudes in children such as persistence. Observation can be structured by using a table format to record 'on-task' and 'off-task' behaviour, such as the checklist of 'persistence' below in Table 12.1.

Table 12.1   Checklist for 'persistence'

Indicators	Observed frequently	Observed sometimes	Not yet observed
Does not give up			
Tries several strategies			
Seeks several solutions			
Other:			

(Johnson & Johnson 1996)

> JOURNAL ENTRY 12.3 *Using a checklist for persistence*
>
> In a group of three assign one member of the group to the role of *checker for understanding*, who has the responsibility for asking other group members to explain the procedures and processes they are using to solve the challenge. Another group member is the *recorder* who uses the 'checklist of persistence' shown in Table 12.1 to record the degree to which the other members of the group are demonstrating persistence. The role of the third group member as *reporter* is to report back to the tutorial about the interaction of the group.
>
> The design brief for the challenge is open-ended and involves textiles as a material (see Figure 12.5).

Imagine you are back in Lilliput, land of the little people. Design and make some Lilliput people and their clothes.

Figure 12.5 Design brief

In the adventure story, on waking, Gulliver met the Emperor of Lilliput who needed a stage with ladders in order to talk to him. Gulliver described this 'person of quality' in the following way:

> *His dress was very plain and simple, and the fashion of it, between the Asiatic and the European: but he had jewels, and a plume on the crest. He held his sword drawn in his hand, to defend himself, if I should happen to break loose; it was almost three inches long, the hilt and scabbard were gold enriched with diamonds. The ladies and courtiers were all most magnificently clad, so that the spot they stood upon, seemed to resemble a petticoat spread on the ground, embroidered with figures of gold and silver* (Swift 1952, p. 13).

OBSERVING FOR SOCIAL SKILLS

Teachers can evaluate children's efforts of working together cooperatively by making anecdotal observations. Social skills can be taught by focusing children's attention on skills such as those in Table 12.2 below.

Table 12.2   Observation of social skills

	When We Work In Groups We
G	Give encouragement
R	Respect Others
O	Stay On Task
U	Use Quiet Voices
P	Participate Actively
S	Stay In Our Group

(Johnson & Johnson 1996, p. 5)

GROUP PROCESSING: ENSURING EVERY GROUP MEMBER RECEIVES POSITIVE FEEDBACK

It is important to give children time to process the effectiveness with which group members work together, and for the group to celebrate their successes; for example, pat one another on the back at the completion of the session. Ways positive feedback can be given to each member of the group can include one member at a time telling the target person one thing that he/she did that helped the group learn or work together effectively. The focus is rotated until all members have received positive feedback. Members can also write a positive comment about each of the other member's participation on a card. The cards are given to each other so that every member will have (in writing) positive feedback from all the other group members.

MULTI-ABILITY TASKS FOR COOPERATIVE LEARNING GROUPS

Gardner (1993) suggests seven possible types of intelligence: linguistic, logical/mathematical, spatial, musical, bodily/kinaesthetic, interpersonal and intrapersonal (see Table 12.3). As most classroom activities only use and develop linguistic and logical mathematical intelligences children who are relatively high in these intelligences tend to always be the helpers in groups, while children who are relatively high in other, neglected intelligences are the ones always receiving help.

Table 12.3 Types of intelligences and their educational implications

Type of intelligence	What people high in this intelligence do well at	Preferred way to learn
Linguistic	Reading and writing, storytelling	Saying, hearing, seeing and writing language
Logical/Mathematical	Logical reasoning, mathematics	Categorising, working with formulae, deductive thinking
Spatial	Puzzles, reading maps and charts, seeing shapes, visualising	Drawing, designing, working with visuals
Musical	Singing, picking up rhythm and sounds	Songs, chants and other materials with rhythm
Bodily/Kinaesthetic	Physical activities	Hands-on, role-play, mime
Interpersonal	Working with others, making friends, helping groups work well	Group activities, interviewing, debates
Intrapersonal	Understanding oneself, metacognition, pursuing one's own interests	Activities with individual accountability, reflective activities

(Adapted from Faggella & Horowitz 1990 by Jacobs et al. 1995, p. 136)

When planning technology tasks we should think about integrating a variety of intelligences into each activity. Multi-ability tasks are more inclusive because they provide opportunities for all children to *give* and receive help, and they encourage each group member to develop in a more well-rounded way (Jacobs et al. 1995, p. 137). For example, by including songs in a technological activity (as suggested in Chapter 6), we are involving musical intelligence.

SUMMARY

In this chapter you have considered comments made by children which indicate that 'making' things is an enjoyable part of technology. You were challenged to make straw towers, little people and their clothes by collaborating in cooperative groups.

When we arrange children in groups to engage in a technological activity we are asking them to investigate the problem or need, generate designs, make the product and evaluate how well it works. If they are to achieve success we must provide prior opportunities for them to develop the social skills and collaborative skills which will enable them to work effectively as a team. Cooperative technological learning therefore encourages children to engage in the design process and use collaborative skills which can be transferred to a variety of situations. The experience of group processing (evaluating how well each member worked in the group) raises the children's awareness of the need to develop and use social skills. Positive feedback from team members increases self-esteem, confidence to risk-take, willingness to tackle problems and take the initiative when managing resources. Therefore cooperative technological learning can contribute to the teaching of these highly desirable skills which form 'enterprise education'.

REFERENCES

Asimov, I. (1980) *The Annotated Gulliver's Travels*, Clarkson N. Potter, Inc, New York.

Davidson, N. (ed) (1990) *Cooperative Learning in Mathematics: A Handbook for Teachers*, Addison-Wesley, Menlo Park, California.

Gardner, H. (1993) *Multiple Intelligences: The Theory in Practice*, Basic Books, New York.

Jacobs, G.M., Lee, G.S. & Ball, J. (1995) *Learning Cooperative Learning via Cooperative Learning. A Sourcebook of Lesson Plans for Teacher Education on Cooperative Learning*, Southeast Asian Ministers of Education Organization (SEAMEO), Regional Language Centre, Singapore.

Johnson, D.W. & Johnson, R.T. (1987) *Learning Together & Alone: Cooperative, Competitive, & Individualistic Learning* (2nd edn), Prentice-Hall, Englewood, New Jersey.

Johnson, D.W. & Johnson, R.T. (1996) *Meaningful and Manageable Assessment Through Cooperative Learning*, Interaction Book Company, Minnesota.

Johnson, R.T. & Johnson, D.W. (April 1997). Notes of workshop in cooperative learning and constructivism conference, RECSAM, Penang, Malaysia.

Kagan, S. (1992) *Cooperative Learning*, Kagan Cooperative Learning, San Juan Capistrano, California.

Klindworth, A. & Jane, B. (1993) 'Towers in technology. Making the most of construction activities', *Investigating Australian Primary & Junior Science Journal*, vol. 9 (2), pp. 6–8.

Slavin, R.E. (1995) *Cooperative Learning: Theory, Research and Practice* (2nd edn), Allyn & Bacon, Boston.

Swift, J. (1952) *Gulliver's Travels*, Aldine Press for J. M. Dent & Sons Ltd, London.

Tytler, R. (1990) 'Raising towers', in *Professional Readiness Study Understanding Science*, Deakin University Manual, Deakin University Press. Burwood.

Van der Kley, M. (1991) *Cooperative Learning: And How to Make it Happen in Your Classroom*, Macprint Printing, New Zealand.

Waterworth, P. & Shepherdson, J. (April 1993) 'Learning to teach cooperatively', in *The Social Educator*: Special conference edition (pp. 44–50).

ACKNOWLEDGMENT

Sally Druit illustrated the *Gulliver's Travels* adventure story.

Chapter Thirteen

Conclusion: what is the magic ingredient?

INTRODUCTION

My earliest memories of designing, making and appraising with materials were in my early childhood years making cubbies. I remember one time when I was about three years old, a delivery truck pulled up and dropped off a huge freezer at our house. Whilst the adults were busy setting up and testing their new acquisition, I was creating the most elaborate cubby with the box! I remember the cosy feeling, the excitement of changing my design, and the thrill of pretending I was a rabbit in my new hutch. (Marilyn)

We used my mother's huge umbrella as the base for our cubby. My brother and I would take the umbrella outside and drape old sheets, blankets and any other item we could find around the edge. The space was special. We planned tea parties, pirate raids and moon walks under our magical umbrella. (Mandy)

Figure 13.1   *The dog and I found this poor bird in the paddock. I made a cubby for it.*

If you mention the word 'cubby', you are likely to find that most people will have childhood memories of making and playing in cubbies. Generally, we remember our cubby building and playing times with great fondness.

JOURNAL ENTRY 13.1   *My experiences of making cubbies*

What are your memories of making and playing in cubbies? Record your memories.
   Note the range of different cubbies and different contexts in which they were created.
   Did you use a range of technological skills or artefacts?
What did you do in the cubbies?
What games did you play?
What play systems did you create?
How did you feel?
Record your ideas and take them to tutorials for discussion.

The designing and making of cubbies is a technological activity. Playing in the cubbies provides a real context in which to appraise the cubby design and construction techniques. Cubbies can be:

- built out of natural materials such as branches or stones;
- constructed from manufactured items such as pine boards or bricks;
- made out of everyday items such as blankets or sheets;
- constructed from found materials such as old boxes or crates; and
- built in discarded items such as old buildings or cars.

As a result, the theme of cubbies lends itself to exploring a range of technological processes, skills and materials.

What is interesting about the theme of cubbies for technology is that the design focus is driven by the children themselves. There is a social purpose for its construction—that is, playing with your friends in a space that is special and usually free of adults!

In thinking back to your childhood and remembering your cubby building experiences, you would have also brought back a range of 'feelings' that were associated with those experiences. If we were to metaphorically bottle those feelings and label them we could then talk about this important effective component in planning for technological learning. Those feelings may include:

- achievement in building a structure;
- connectedness with self and others;
- cosiness and closeness;
- security;
- privacy;
- inner warmth;
- enjoyment;

- self-direction;
- self-motivation; and
- secrecy.

This 'connected awakening and achievement' type feeling that is associated with cubby building is an important and magic ingredient for fostering learning opportunities. As teachers we should consciously plan to include this magic ingredient for all the technological learning environments that we create for children. How can we ensure that the technological learning challenges we organise for children take account of this magic ingredient?

JOURNAL ENTRY 13.2 *Reflecting on the approaches to teaching technology*

In Part 2 of the book you were encouraged to explore a range of ways of teaching technology to young children. You were asked to consider the science–technology relationship; a symbiotic approach; a process approach; and an ecological approach. For each approach think about the following:

- Do you believe the magic ingredient of 'connected awakening and achievement' feelings will emerge in the approaches discussed?
- What was the role of the teacher in teaching technology?
- How do you feel about the different approaches to teaching and learning in technology?
- In each approach what assumptions are being made about: children's learning? The nature of technology?
- Who asks the technological questions—the children, the adults, the community?
- Will the technological experience connect with the child?
- Is there a social purpose for the child in designing, making and appraising?

Discuss and compare your analysis with others who have completed this experience.

The social purpose of technology was considered throughout this book. If we consider the theme of cubbies it is possible to see how this connects with babies and toddlers, primary school-aged children and even with adults. Three examples which focus on cubby building follow. Note the differences and the similarities in the technological activities undertaken by the various age groups.

CUBBIES FOR INFANTS AND TODDLERS

Kay and Carolyn lead the planning and teaching of this unit.

Kay: 'Building and playing in cubbies is an important part of our infant program. Sometimes we set the scene by providing the stimulus. For example, we made a basic frame out of quadro and the babies helped us cover it in a range of textured fabrics

continues...

(mostly bedspreads). The infants then moved about in the space—going in and out; crouching down, stretching up; moving the fabric about to suit their exploration. Many of the babies found small spaces or holes which they used to play peek-a-boo.'

Carolyn: 'At other times the babies will take the lead. Yesterday, Julia took the foam mattress and tried to move it around herself. We saw what she was trying to do and immediately assisted her. She motioned how she wanted the foam pieces to be placed and I acted as a builder's labourer physically moving things to suit her design.'

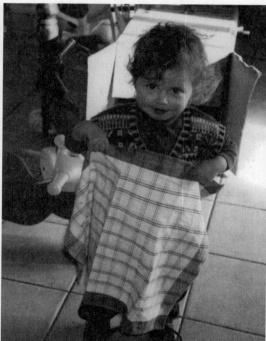

Figure 13.2   Peek-a-boo in the cubbies

CUBBIES FOR PRIMARY-AGED CHILDREN

Mandy is the teacher of 19 five and six year olds in a small rural school. She has noticed that during lunchtime and recess the children play almost exclusively under the branches of a cluster of trees behind the classrooms.

Mandy: 'The children have developed a range of cubbies under the trees. Although they have not used many materials, their activity demonstrates all the characteristics of cubby playing. As a result, I decided to incorporate their play and interest in cubbies into a technology unit.'

continues...

The children came back to school in Term 2 to be greeted by the opportunity to make and play in a range of cubbies in the classroom. The children drew pictures of the cubbies they made, wrote stories (with adults scribing) what they did in the cubbies and had extensive discussions about the best design or most durable style of cubby that they had made. Mandy followed up the free play sessions with the reading of a range of books on making and playing in cubbies. The children decided to begin planning a 'super cubby' that would be 'really cool' to look at and play in.

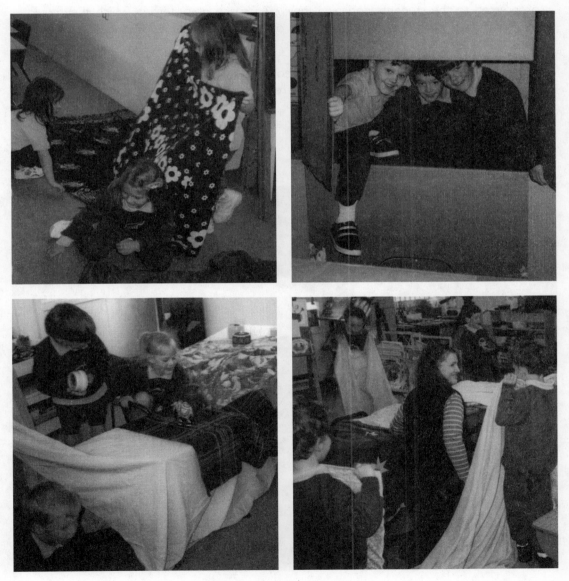

Figure 13.3 Let's make a 'cool' cubby!

CUBBIES FOR ADULTS

Second year preservice teachers about to embark on the study of technology education arrived at their tutorial room to find a range of cubbies awaiting them. They were told to 'play in the cubbies' and to 'talk to each other about their childhood memories of cubby building and playing'. The student teachers roamed around the room exploring, talking and reliving fond memories from the past. The students recorded their ideas in their journal and then were asked to think about what space they now enjoy being in. This progressed to the students designing their own space or system in which they could relax and enjoy the semester work ahead of them. The students produced a range of designs in which:

- a system of refreshments was organised;
- specifically arranged musical pieces were heard;
- beanbag-style chairs were sewn; and
- colourful posters, pot plants and other artefacts were incorporated into the space.

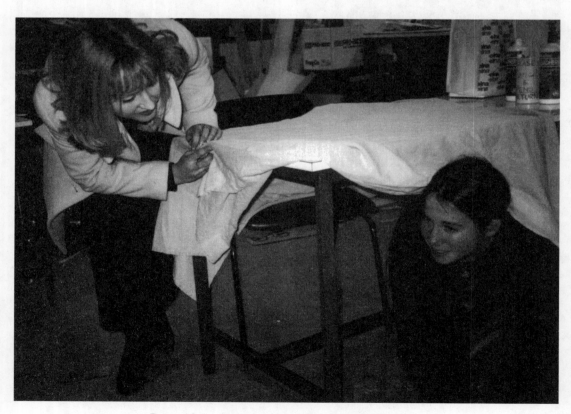

Figure 13.4   Designing a comfortable semester space

> JOURNAL ENTRY 13.3 *Deconstructing cubbies*
>
> What did you notice about the technological learning that was occurring in each context described above?
>
> What was the same? What was different?
> What was the role of the adult in each instance?
> Were the technological learning outcomes the same for each age group?
> Record your responses and take them along to tutorials for sharing.

In deconstructing the 'construction of cubbies' you are effectively engaging in program evaluation. Journal entry 13.3 asked you to evaluate the three cubby programs detailed earlier. In analysing the programs you were making decisions about the teaching–learning process as well as the learning outcomes for children. You would have noted that the learning outcomes for each age group were effectively the same—developing children's technological capability. However, the level of skill and understanding was different. You would have also noted that the teaching approach used by each teacher was also the same—a process approach. The children were designing, making and appraising with materials and systems. The evaluation of teaching programs also involves a third step—self-evaluation. In this next section, the principles of program evaluation are discussed.

EVALUATING TECHNOLOGY PROGRAMS

As a teacher you will make decisions about the program you have developed (as you did in journal entries 13.2 and 13.3) and how successfully it was implemented. This latter task is generally referred to as an 'evaluation'. Unlike the assessment and profiling of children's understandings, evaluation is a broader concept involving a critical review of all elements of a teaching program. The three areas normally evaluated include:

1. self-evaluation;
2. evaluation of the approach taken; and
3. evaluation of learning outcomes.

Self-evaluation is an important part of evaluating a technology program. Teachers should ask themselves:

- How did I feel about the technology teaching approach I used?
- Did the topic lend itself to stimulating and challenging questions emerging from the children?
- Did the management of the children and the resources influence the program outcomes?
- What did I learn about technology teaching and children's thinking?

The *evaluation of the approach* taken is important. Consideration should be given to the effectiveness of the teaching–learning process. For example:

- Was I aware of the understandings children brought with them?
- Were children the centre of the learning process?
- Were children actively participating in the learning process?
- Were girls participating and achieving equally with boys?
- Was I using a variety of roles and strategies?
- Were the learning experiences relevant to everyday life?
- Were safe practices being adhered to by all?
- Did children ask their own technological questions?
- Were the children encouraged to plan their own design briefs?
- Were children encouraged to play with their ideas?
- What discourses were operating?
- Did the girls have agency?
- Were multiple world views considered and valued?
- Are appropriate technologies being considered?
- Are ecological implications being considered?

It is also important to consider *the evaluation of the learning outcomes*. Teachers profile individuals and make professional decisions on the learning outcomes that need to be planned. In Chapter 9 a range of ways of assessing children's learning was featured. For example, assessing learning could occur through:

- observations of children at work;
- work samples;
- teacher–child conference;
- group discussion;
- self-reporting;
- checklists;
- performances (role-plays, debates, drama, songs, poetry);
- audio-recording of small group and whole group discussions; and
- profiling students.

It was also shown that a range of contexts for assessing children's learning could be constructed. In particular, play featured as an important element in organising children to work at their optimum level. In play children are:

- motivated;
- engaged in a purposeful and relevant task;
- working to capacity;
- making connections, working abstractly, communicating and exploring tentative ideas; and
- talking 'out loud' their ideas and thinking processes.

Assessing children's learning allows a teacher to determine whether or not the learning outcomes were achieved. As indicated earlier, student assessment is only one of the variables that need to be considered when evaluating a technology program.

Effective evaluation of the technology program is possible when a teacher self-evaluates, analyses the approach adopted in relation to its implications for children's learning, and when the overall program outcomes are considered in relation to what children have *actually* learned.

SELF-EVALUATION

The content of this book was designed to be a cognitive journey. At this point you should have well formed views on how you will *program*, *teach* and *evaluate* technology education. That is, you would have *developed your own approach to the teaching of technology* (which is the subtitle of this book).

In Part 1 of the book you were invited to think through a range of issues associated with planning for children's technological learning. You were asked to consider the types of technological experiences children have in their home. The home experiences were considered as useful starting points for teachers when planning technological tasks which build upon their existing technological capabilities. In considering the home context, Chapter 3 moved beyond a Western view of technology and investigated the notion of multiple world views in curriculum design and implementation. Although cultural interpretations are discussed throughout the book, this chapter challenged the assumption of a Western world view of technology and curriculum design.

The journal has allowed you to map your thinking about what technology is as you have progressed through the book. In particular, note the responses you made in Chapter 1.

JOURNAL ENTRY 13.4 *Reflecting on your own learning*

How do you react now to your earlier journal entries?
How has your thinking changed? Record those areas in which your thinking has developed.
What do you still need to know more about? What knowledge and skills would you like to further develop in technology education?
Why are these important to you?
Share your thinking with your colleagues in tutorials.

SUMMARY

Designing, implementing and appraising your own technology learning and teaching program is clearly a technological task for you as a teacher. The magic ingredient for you, of course, is not just the learning outcomes for children, but the love for learning that can be fostered. This enthusiasm for technological learning is captured well in the following keynote address given by a secondary student who reflects upon her technological learning experiences. Georgina demonstrates the importance of the home, the organisation of technology within school and what it means to be a girl engaging in technology. If we can graduate children at the end of primary school with Georgina's enthusiasm and skill in technology, then we would have successfully brought together *teaching* and *learning* in ways which reflect the magic we identified in cubby building.

> 'When Chris Thomson first asked me to do this I was a little apprehensive at the thought of standing up here and talking to teachers when usually it's the other way round. I wondered why on earth would a group of experts in their field want to hear from me, a Year 7 student?
>
> I received my first set of tools and a pair of overalls from my Dad when I was two years of age. Much to Mum's horror they weren't plastic toys but the real kind. My mother had visions of me ordering four Big Macs with only two fingers!
>
> I quickly learned that if I was to succeed at technology I had to learn some skills. This fact was brought home to me the first time the hammer 'bit' my fingers. I was taught to hold tools correctly. Dad would say: "don't choke the hammer!" Now, I knew that I wasn't to choke the puppy but how could I choke a hammer? I realised that technology not only had its own set of skills but its own language as well.
>
> Now I don't want to give the impression that the male species dominates technology in our house. My Mum is quite a handyperson too. She has made the chest in my Mum and Dad's bedroom that Dad now frequently stuffs his old jumpers into. I also remember my Mum and my Nanna making all of my clothes when I was younger. Many times I was used as a tailor's dummy, having measuring tapes draped over me and pins and needles stuck into everywhere possible. It was here that at an early age I learnt how things were manufactured and that they didn't just end up on a shelf in a store. In all of these situations in my early childhood I unconsciously became aware of the fundamentals of technology—investigate, design, make and evaluate.
>
> My formal introduction to technology began in kindergarten where they had a woodwork bench and here I experimented on various ways of joining wood. I found that I could use string, wood, plastic, paper or anything else I could lay my hands on to do technology.
>
> When I started primary school, the technology I did was basically paper and cardboard engineering but things got a little bit more interesting when a Japanese teacher joined the teaching staff. He started a lot of lunchtime groups and one of them was the origami group. Every Monday we would meet in the library and make things like origami purses, miniature warrior helmets or even paper drinking cups.
>
> Because of my early start to technology and my parents' background I had a head start when it came to school projects. This sometimes led to accusations, some even from teachers, that the work submitted was not my own. But people didn't realise that at home I had an extensive range of tools, equipment and resources at my disposal. We could ask some questions perhaps. Was it my fault that my parents have a lot of tools and resources? Was I supposed to act dumb and not try my best? Was it not the responsibility of the school to make sure I had access to these resources and to encourage me to develop and extend an area of expertise?
>
> *continues...*

When entering Year 7, I was amazed at the range of options open to me. I also found that the school valued all aspects of technology not just textiles and woodwork. The equipment and resources in the school were more in line with what I had become used to in my parents' workshop but there is a wood lathe which we don't have at home and I just can't wait to be let loose on it!

My mother's experiences of technology were vastly different. In primary school, on a Wednesday afternoon, all of the girls from Grades 3 to 6 were taken by a female teacher for needlework and knitting while the boys were taken by a male teacher to do leatherwork, basket-making and gardening. In secondary school the girls did needlework and cookery to prepare them for their 'domestic role in life'. However, the boys studied woodwork, metalwork and mechanical drawing to prepare them for their role as the 'Captains of Industry'. And never the twain shall meet. Evidently no one ever thought that girls might enjoy woodwork and metalwork and actually be good at it, or that boys might enjoy textiles and cookery. Thank goodness a generation of educators have begun to think differently.

So far as being a student, what have I got out of technology? Enjoyment. I certainly enjoy my time in the technology area at school and I enjoy technology-based activities. I had a first among my girlfriends when I had a technology party theme for my birthday party. We made model solid fuel rockets and had a great time!

I think that because technology is a practically-based subject and the emphasis is on investigating, designing, making and testing this provides an enjoyable and often refreshingly different way to learn.

At this point in my life I am unable to say what changes will occur in the world or what my future is. But I do know that I will need to work, live and grow in that future world.

So to put the shoe on the other foot for a moment (that is my shoe on your foot), I set the following challenge to you, the designers and teachers of technology education:

You must ensure that my needs and those of thousands of students like me are given the highest priority and serious consideration, and that the technology curriculum that you provide and the way in which you provide it prepares me to meet the challenges of the next millennium.'

(Georgina Seddon, Year 7)

ACKNOWLEDGMENT

Many thanks to the University of Canberra preservice students, Mandy Perry and her children from Hall Primary, and Kay Howell and Carolyn Bennett and the infants from Wattle Childcare.

Index

aboriginal technology
   cycad extractive technology 48
   didgeridoo 48
   humpy 46–47
   mia mia shelter 44–45
   spear 38
   world view 37–38, 135
approaches to technology teaching
   symbiotic approach 81–94
   de Vries' classification 40
   ecological approach 111–122, 133–135
   process approach 95–110, 126–129
appropriate technology 17–19, 115–117
assessment
   observations 139, 186
   profiling children 150–151
bees 135
biotechnology 73–74
bird feeder 69
border crossing 43
bread making 73–74
bridges 179
canning 127
children's questions 57, 89, 153–160
clockwork radio 111–114
coconut 133
computers 4, 100
content
   design 96, 139–149, 153–160
   information 159, 164–165, 168–170
   materials 86, 158–159, 162, 181
   systems 50, 129, 157–158, 159
co-operative learning
   basic elements of co-operative groups 183–184
   co-operative skills 180
   jigsaw models 184–185
   multi-ability tasks 188
   observing 186, 188
   STAD 185
   persistence 187
   processing group functioning 182–188
   social skills 188
curriculum
   Aboriginal perspective 38–39
   D&T 40, 60, 132
   Eastern perspective 50
   integrated 81 130–131
   South Africa
   STS 59–60
   Western perspective 40
culture 9–11, 16, 17–19, 27, 37–53
   Maori tools 42
design brief 40–41, 63, 72, 85, 181, 187
DMA 96 139–149, 153–160
dreaming stories 48–49
ecodesign 155–160
evaluation 197–200
FPTs 132–134
gender 11–16, 161–174

discourses 172–174
Gullivers Travels 180–181, 187
icecream making 58–59
IDEAs 132–133
intelligence types 188–189
inventors 112–113
learning environment 91–92
Lilliput challenge 180–183
outcomes based education 149
planning 84, 96
   environment 105, 139–140, 191–200
   framework 156, 159
   interactions 161–167
   routines 105
prior knowledge 85
quality technology education 132
real life contexts 75–77
school-community links 118–119
science knowledge 74
science-technology relationship
   materialist view 61–67
   demarcationist view 75–77
   TAS view 67–70
   interactionist view 70–73
   symbiotic view 73–75, 81
slater 88–91
slug 89
small animal catching device and container 85–87
social constructivist perspective 91–92
S&T centre 119–120
technology
   appropriate technology 17–19, 115–117
   child-birth technology 11–17
   child-care technology 26–35, 98–100, 140, 144–147
   children's experiences 23–35, 191–193
   definition of 4–9, 106–109
   home technologies 6–8, 23–35
   history 9–11
   knowledge 74
   perceptions of 82, 118–119
   skills 74
   social shaping of technology 11–17
   status 82–83
   sustainable 115
   technology transfer 18–19
textiles 50–51, 187
the teaching of
   infants and toddlers 26–35, 98–100, 109, 193–194
   out-of school-hours care 106
   preschool children 107, 139, 141–143, 144–147, 158, 165, 167
   school children 100, 107, 148, 157, 194–196
thinking styles 126, 131
tinkering 61, 70, 170–171
torch 70
toys and gadgets 62–66
values 39
world views 43
Yowie 123–124